Heterostructural
Interface Modelling

David J. Fisher

Copyright © 2019 by the author

Published by **Materials Research Forum LLC**
Millersville, PA 17551, USA

Published as part of the book series
Materials Research Foundations
Volume 60 (2019)
ISSN 2471-8890 (Print)
ISSN 2471-8904 (Online)

Print ISBN 978-1-64490-046-8
ePDF ISBN 978-1-64490-047-5

Distributed worldwide by

Materials Research Forum LLC
105 Springdale Lane
Millersville, PA 17551
USA
http://www.mrforum.com

Printed in the United States of America
10 9 8 7 6 5 4 3 2 1

Table of Contents

Heterostructural Interface Modelling

Just as small dimensions are becoming more and more important with regard to nanocrystalline metals, and with regard to free-standing atomically-thin materials such as graphene and its analogues, so too are what one might term the zero-dimensional components which are a feature of all materials: their surfaces and interfaces.

Sometimes seen as flaws, and sometimes as essential features, they are increasingly studied in their own right because what happens at them is often the key factor which will decide, for example, whether a given electronic device will function as intended. In the case of semiconductor electronics alone, the choice of the means of connection must take account of phenomena such as electron-band alignment, the transport of electrical charge, mechanical straining, chemical bonding and the incidence of point defects and impurities.

At the same time, the materials whose surfaces have to be brought together at some stage have become increasing exotic, and include perovskites, spinels and olivines. As a concrete example, there is the possibility of introducing a new class of solid-state lithium-ion conductor having the spinel structure. Such compounds might be coupled to spinel-type electrode materials so as to produce a lattice-matched solid-state device having a low interfacial resistance. In one experiment[1], marked differences in the Li^+ diffusional properties were observed, depending upon the composition, lithium content and cationic distribution. Local Li^+ hopping in the spinel was associated with a low (0.35eV) activation energy.

The modelling and prediction of interfacial phenomena are consequently becoming increasingly important. When the abutting materials have the same type of structure, an atomic model can be easily constructed via various rotations of the unit-cell. This has long been a staple technique in metallurgy. A striking example of interface-matching, which is sometimes demonstrated physically at conferences on cast-iron, is to draw an hexagonal pattern on transparent plastic sheet and then to roll, or otherwise manoeuvre, the sheet so as to bring the hexagons at various points of the sheet into exact correspondence. The point of this is that various forms of filamentary growth of graphite from liquid or gaseous phases have structures which can be interpreted in terms of curved basal sheets of graphite that are arranged in the form of cone-helices. In these arrangements, successive spiral basal layers are rotated relative to one another. Usually only those geometries are permitted for which the rotation is a multiple of 60°. This restores the graphite basal stacking in the c-axis direction. However, a rotation of 21.8° is also acceptable. It is one of the optimum coincidence configurations for c-axis rotation faults in graphite. And it also seems to explain the spherulitic form of graphite in cast iron, if one imagines the spheres to be made of conical particles having an included angle which is governed by the above sheet overlap.

Although such historical roots can offer an initial clue as to the likely structure of an interface, real materials can possess properties which modify or invalidate the expected structure. There is already a fundamental incompatibility involved in the basis of crystallography itself: that is, the structures which small isolated groups of atoms adopt in order to satisfy a minimum-energy criterion will not generally, if ever, satisfy the criterion of uniform space-filling which is demanded of an entire macroscopic crystal. One can therefore well expect thermodynamic factors and defect-formation to compete with straightforward lattice-matching criteria. This is especially true of dislocations, which frequently compete with pure geometrical lattice-matching in the minimisation of interface stresses due to misfit.

Interfaces between different parts of the same material have always been important with regard to grain boundaries, twins, stacking faults and more complicated defects. Attention is directed here however towards interfaces which bring together two materials possessing different crystal structures. This has also been an important area of study in the past with regard to epitaxial deposition, precipitate-strengthening and composite design.

The basic incompatibility, probably even of the unit-cell dimensions, immediately conjures up interesting mathematical problems but these are ones which have, fortunately, often been solved already by mathematicians, if only in a very abstract sense. More practical approaches, it will be found, include coincidence site lattice theory (originally developed in order to treat the homogeneous interface case) which analyses the fundamental atomic arrangements. That is, for certain misorientation angles there may exist more coinciding lattice sites, in addition to that at the axis of rotation. Such configurations are assumed to be more stable than others, and this seems to be usually true of metals but, in an ionic material, the oppositely-charged ions tend to favour propinquity. That is, oppositely charged ions may prefer to minimise their separation.

In the past, the successful prediction of an interface structure was an end in itself. Analysis tends to be deepened nowadays in order to determine relevant interface properties given that, unlike previously, there is now a much greater probability of being able to tailor those properties. The total interface energy can be deduced from the area of the interface, the chemical potentials of the various atomic species (which include non-stoichiometry and impurity effects) and possibly any mechanical strain energy. The interface energy tends to be of the order of $0.3 eV/\mathring{A}^2$. Strain can have the useful effect of increasing the carrier mobility, as when silicon is intentionally strained for that purpose.

Lattice-mismatch can increase the ionic conductivity along strained heterostructural interfaces by orders of magnitude[2]. An exact model for the depth evolution of such ionic conductivity improvement by interfacial lattice strain permits the exact evaluation of the correlation, between lattice strain and ionic conductivity increases in an isotropic orthorhombic lattice, by taking account of factors such as the Young's modulus and Poisson ratio.

The crystal structure and symmetry of epitaxial thin-film systems deviate from the bulk properties due to epitaxial strain and interfacial coupling[3]. The crystal symmetry of perovskites is described by rotations of six-fold coordinated transition-metal oxygen octahedra. These are changed by the proximity of interfaces. The local oxygen-octahedral coupling at perovskite heterostructural interfaces markedly affects the domain structure and symmetry of the epitaxial films. The interface can thus be tailored by choosing suitable substrate/buffer/film combinations. Such apparently local combinations unexpectedly lead to structural changes which affect the entire thickness of the film.

A recently proposed method[4] for identifying those compatible material combinations which offer mechanically and electronically desirable properties is based upon a procedure in which preliminary matching of lattice constants ensures the choice of interfaces exhibiting minimal epitaxial strain and thus maximum mechanical and chemical stability. Absolute electron energies are then matched in order to construct energy-band alignment diagrams.

Many interfacial properties are strongly related to the geometry and chemistry of the interface and therefore correct modelling of the interface at the atomic scale is essential. Such 'pattern-matching' has its roots in very early abstract mathematical thought.

The underlying problem involved here, the geometrical relationships between lattices, has a long history; going all the way back to Newton's study of the so-called kissing-number of balls, Kepler's study of ball-packing densities, Lagrange's deduction of the maximum lattice-packing density of discs and Gauss' determination of the densest lattice-ball packing.

Hermann Minkowski, better known in connection with the special theory of relativity, is credited with providing (1891) the most fundamental theorem in the field known as the geometry of numbers[5]. This states that, if S is an n-dimensional centrally symmetrical convex body centered at the origin and if the n-dimensional volume is such that $V_S \geq 2^n$, then S contains an integral point which is distinct from the origin. Apart from the present context of crystallography, the geometry of numbers has numerous applications throughout number theory: algebraic, transcendental and Diophantine. It also plays an important role in geometry (discrete, convex and computational) and in coding theory.

Minkowski later (1896) pointed out that every lattice-square tiling of the plane contains two squares which are joined along an entire edge and, moreover, that every lattice-cube occupation of three-dimensional space contains two cubes which are joined over an entire facet. Although not relevant to the present work, which essentially concerns overlayers of differing two-dimensional lattices, it is interesting to note that Furtwängler's conjecture (later disproved for higher dimensions), that a twin exists in every multiple lattice-cube space-filling, still applies to the two-dimensional case.

Minkowski's fundamental result in the geometry of numbers was apparently inspired by the Diophantine problem of three integers. Given a, b, c, with a > 0, such that $ac-b^2 = 1$, prove that $ax^2+2bxy+cy^2 = 1$ always has an integral solution. Minkowski used an Euclidean coordinate system in the plane, with $ax_1x_2 + bx_1y_2 + bx_2y_1 + cy_1y_2$ as a scalar product, and took d to be the Euclidean distance to the origin of the next integral lattice point (x,y). Then $ax^2 + 2bxy + cy^2 = d^2$, and it remains to show that $d^2 = 1$. To do this, one centers a sphere with radius, d/2, on each lattice point. This leads to a sphere-packing with a density of $\pi d^2/4$. Observing then that d^2 is an integer which is smaller than $4/\pi$ thus completes the proof. It also illustrates the close connection between lattice-packings and Diophantine problems, with lattice packings corresponding to positive definite quadratic forms.

Software was first developed 2 decades ago for making exhaustive searches of crystal-structure databases in order to identify all orientations of all known compounds, and match the lattice of any plane of any desired epitaxial deposit. Programs could screen for crystal-symmetry, orientation, material type and mismatch limits. Searches could nevertheless still yield hundreds or thousands of potential substrates, and require further screening using criteria such as: feasibility of synthesis, chemical/physical compatibility, etc. In some cases, such as that of protein-crystal growth on inorganic substrates, the matching unit-cell of the material to be deposited can be far larger than that of any candidate substrate.

Coincidence-Site Lattice Theory

This is the earliest[6] method for analysing the interfaces between crystals in contact. Complete rotation of a lattice obviously brings it back to complete self-coincidence, but partial coincidences can appear at intermediate degrees of rotation about the axis. The crystal lattices which are related by such an intermediate angular rotation share common sites which are found to be located on a single lattice having a greater cell-size: the coincidence-site lattice. Such lattices have long been of interest in connection with recrystallization textures, grain-boundary migration and grain-boundary structure.

The misorientation relationship between two crystals can be defined by an axis and an angle. In the case of the [100] axis of the cubic system a lattice point is chosen to be the origin and a square cell is constructed on each line which connects that point to the nearest 'line-of-sight' point. Each of the resultant cells is square, and its size is greater than that of the original unit-cell, thus producing the coincidence-site lattice. The ratio of the areas of the cells is x^2+y^2, where x and y are the coordinates of the lattice-point which is connected to the origin. It is also the so-called multiplicity, Σ, of the coincidence site lattice: defined to be the reciprocal of the density of points-in-common. Note that the case of [110] is less obvious, because there are no squares and the area of a rectangle based upon a line joining the origin to the nearest line-of-sight point has to be used instead. A rotation of π around [hkl] in the cubic system generally gives rise to a coincidence site lattice of $\Sigma = h^2 + k^2 + l^2$, if the latter expression is odd, or $\Sigma = (h^2 + k^2 + l^2)/2$ if the expression is even.

In a key paper[7], the general coincidence site and displacement shift complete lattice model for internal boundaries in a crystalline material was expounded. The model was applicable to grain boundaries in cubic and non-cubic materials, and to interphase boundaries. It was described as being 'fit-misfit', where the 'fit' regions were patches where partial lattice matching across the boundary occurred, the 'misfit' regions were boundary line defects of dislocation/boundary-step type. The degree of 'fit' was then determined in effect by the size of a suitable coincidence site lattice formed by the lattices adjacent to the boundary. The Burgers vectors of the line defects meanwhile were vectors of a suitable displacement shift complete lattice which was also made up from the lattices adjacent to the boundary. The vectors which described the nature of the step were defined by the framework of the displacement shift complete lattice.

High-resolution electron microscopic studies of the grain-boundary structures of face-centerd cubic materials such as aluminium, TiC and silicon indeed showed[8] that they all consisted of arrays of lattice and displacement shift complete dislocations. Their densities varied continuously with misorientation, regardless of the type of atomic bonding. It was surmised that the elastic energy of the boundary lattice dislocations accounted for most of the overall boundary energy.

A three dimensional near-coincidence site lattice model, based upon coincidence site lattice theory, has been proposed[9] in which coincident atoms are determined by computer-based atomic-matching. As a test of the method, it was compared with 0-lattice theory for predicting preferred orientational relationships in a body-centerd cubic material in contact with a face-centerd cubic one. Good agreement was obtained. When applied to an $(\alpha+\beta)$ two-phase titanium alloy, the proposed method could predict the observed α/β interface structure.

The coincidence site lattice model was very successful in explaining grain-boundary structures in cubic materials involving metallic, ionic or covalent bonding. Its use was more complicated in the case of non-cubic materials and phase boundaries, where the assumed conditions required for lattice coincidence did not always exist. This problem could be circumvented[10] sometimes by applying a small strain to the crystal lattice, leading to the concept of a *constrained* coincidence site lattice. This modification had far-reaching effects, in that it could no longer be assumed that the boundary energy was a minimum at the exact coincidence orientation. The Σ-values could moreover be odd or even, and vary with misorientation angle. The Σ-values at a triple junction did not even have to obey the so-called quotient rule.

These results are related to the much more abstract field of tessellation theory, and the latter shows that a square cell cannot be constructed on an hexagonal lattice, nor an hexagonal cell on a square lattice. This clearly does not bode well for the matching of interfaces between differently-structured materials.

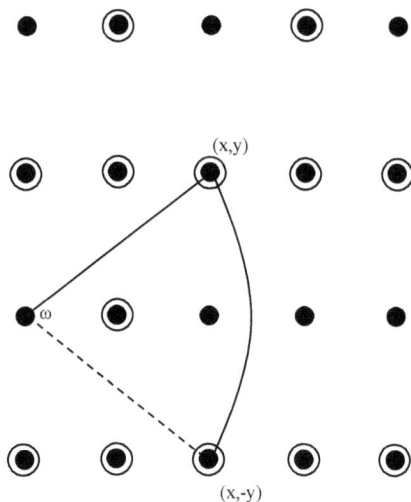

Figure 1. Construction of a coincidence-site lattice in a rectangular lattice
Circled points are 'line-of-sight' points

This led to a search for a generating-function for coincidence-site lattices. In a two-dimensional rectangular lattice (figure 1) with a unit-cell axial ratio of r, a rotation of lattice-point (x,-y) into (x,y) produces a coincidence site lattice as the row containing (x,-y), and the row perpendicular to it, coincide; the angle of rotation being $2\tan^{-1}(y/x)r$. The areal ratio of the two cells is then $x^2 + ry^2$, and the multiplicity of the coincidence site lattice is either Σ or a sub-multiple of it if there are further coincidences within the larger cell.

It is then possible to identify such rectangular lattices on any plane, (hkl), in the cubic system, with [$\underline{k^2+l^2}$, hk, hl̄] and [0lk̄] being perpendicular directions in that plane. The latter directions can be used to construct a cell having the axial ratio, $\sqrt{(h^2+k^2+l^2)}$, and it is easy to see from the area of the cell that there are (k^2+l^2-1) atoms in the unit cell. Each of these is itself the origin of similar displaced rectangular cells.

Because Σ takes only odd values in the cubic system, it is necessary to divide even values of Σ by multiples of 2 in order to obtain the correct multiplicity. Using the generating function, $\Sigma = x^2+(h^2+k^2+l^2)y^2$, acceptable values are obtained by systematically inserting integer x and y values. The above 'line-of-sight' condition requires those x and y integers to be relatively prime; in turn implying that points of the form, (nx, ny), can be ignored because they all give the same coincidence site lattice point: (x,y). In the case of the [111] axis of a face-centerd cubic lattice, where the x and y choices correspond to the [1̄1̄0] and [1̄1̄2] axes, respectively, the coincidence lattice points are given by table 1.

Displacement Shift Complete Lattice

Although the original coincidence-site lattice concept was, and is, very useful, the later use of then-new techniques such as high-resolution electron microscopy revealed that, especially in the case of grain-boundary structures, the relative orientation between two neighboring grains sometimes deviated, by a few degrees, from the coincidence site lattice: the discrepancies being compensated by the formation of dislocation arrays. The displacement shift complete lattice was therefore introduced. This is a lattice which represents the displacement of one crystal with respect to another. The Burgers vectors of grain-boundary dislocations are usually vectors of the displacement shift complete lattice.

In some cases, such as body-centered tetragonal structures, the lattice parameters differ from those of the originally considered cubic structures due to the change in the c-axis ratio. Because it is necessary to have rational values of a^2, b^2 and c^2, an example of the importance of the geometry-of-numbers, the actual lattice parameters have to be rounded-off to the nearest rational value. This led to the concept of the constrained coincidence site lattice, which is important for determining the grain-boundary structures of materials with irrational axial ratios. For example, the grain-boundary structure of $BaTiO_3$ has a multiplicity of 16. This Σ-value, and the angular relationship between two grains, do not match grain-boundary structures which are predicted by the coincidence site lattice of a cubic material. The oxygen atoms of the perovskite structure take up body-centered tetragonal positions and $a^2 : b^2 : c^2$ can be set equal to 1:1:2 by using the constrained coincidence site lattice concept.

Table 1. Coincidence site lattice relationships for [111] in a face-centerd cubic crystal

$$\Sigma = x^2 + 3y^2, \ r = 2\tan^{-1}(y/x)\sqrt{3}$$

x	y	Σ	r(°)
0	1	3	180
1	1	1	120
2	1	7	81.8
3	1	3	60
4	1	19	46.8
5	1	7	38.2
6	1	39	32.2
1	0	1	0
3	1	3	60
3	2	21	98.2
1	1	1	120
3	4	57	133.2
3	5	21	141.8
1	2	13	147.8

Down the years, the displacement shift complete lattice method has been applied to a wide range of situations involving interfaces of various types.

Basal and rhombohedral annealing twins in polycrystalline alumina ceramics were shown[11] to arise from misorientations, and to have Burgers vectors arising from the displacement shift complete lattice; the basal twin being unique to the rhombohedral system. The rhombohedral twin was found to be similar to the $\Sigma = 7$ <210> boundary in hexagonal metals.

When grain-boundary sliding was investigated[12] with regard to 5 types of orientation-controlled bicrystalline specimens of aluminium, a displacement shift complete dislocation analysis showed that the step-vector, in coincidence with a low-index plane of one of the component crystals, controlled the selection of grain-boundary dislocations.

Twinning steps, associated with a twinning dislocation at a (110) twin boundary of $YBa_2Cu_3O_7$, were suggested[13] to comprise 2 types of twin-boundary: one involving a lattice translation along the boundary in the case of fully-oxygenated material, and the other – involving no lattice translation – in the case of oxygen-deficient material. Dislocations appeared at points where the two types of twin boundary merged. The geometries of the twin boundary and twinning dislocation were analyzed using the coincidence site lattice and displacement shift complete lattice concepts.

Molecular dynamics simulations at constant temperature and stress were used[14] to study $\Sigma = 5$ (310) tilt grain boundaries in NaCl, and several metastable core structures were identified. The differences in free energy among these structures were so small that two of them, related by their small displacement shift complete vectors, co-existed at high temperatures.

It was later shown[15] that the displacement shift complete lattice was not required in order to define the Burgers vector which was associated with an interfacial dislocation. In the case of silicon and germanium [011] perfect or imperfect bi-crystals, the grain-boundary structure consisted of a limited number of atomistic structural units. Such a structural unit could be considered to be the core of a grain-boundary dislocation. The limited number of structural units then greatly restricted the number of possible Burgers vectors, as compared with the number of displacement shift complete lattice vectors. This work provided a link between the geometrical and energetic approaches to interface analysis.

The structures of intrinsic ledges at interphase boundaries were explained[16] in terms of the displacement shift complete lattice concept. The distribution of structural ledges could be predicted if the spacing-difference between parallel matrix and product planes was treated as being a measure of the relaxed coincidence condition. A small rotation, away from low-index planar parallelism, introduced interfacial dislocations which cancelled the spacing difference and resulted in a lattice invariant line. The Burgers vectors which were associated with structural ledges and misfit-compensating ledges were displacement shift complete lattice vectors.

Large-angle convergent-beam electron diffraction was used[17] to analyse secondary dislocations in the $\Sigma3$ and $\Sigma9$ grain boundaries of silicon. Upon selecting reflections from crystal planes which were common to the adjoining grains, the images became insensitive to the boundaries, except where dislocations were present. The dislocations could be analysed assuming integer values of g•b, implying that the Burgers vectors were displacement shift complete lattice vectors. In the case of both $\Sigma 3$ and $\Sigma 9$ boundaries, displacement shift dislocations were identified that were specific to those boundaries.

A mean field equation for a [001] twist grain boundary was derived[18] by using coincidence-site lattice symmetry, the displacement shift complete lattice and 2 order parameters. One of them corresponded to the direction of the coincidence-site super-cell lattice vector and one to the [001] normal to the boundary. The solution described the decay-lengths and amplitudes around a grain boundary or free surface and revealed some additional length-scales.

Study of a $\Sigma = 11$ {332} nickel bi-crystal, grown by solidification, showed[19] that the observed grain-boundary atomic structure depended upon the sample's position in the bicrystal. All of the grain-boundary defects could be characterized, in that all of their Burgers vectors belonged to the displacement shift complete lattice.

Coincidence-site and displacement shift complete lattices of triple junctions were analyzed[20] in order to show that, for some triple junctions, no trapped residual triple-junction dislocation is required geometrically for dislocation-transmission between adjoining grain boundaries. In other triple junctions, a residual dislocation sometimes cannot leave the triple junction, for a grain boundary, without generating a stacking-fault type of defect.

The $\Sigma = 9$ [$0\bar{1}1$]/(122) symmetrical tilt grain boundary of copper was analyzed[21], showing that mirror and glide-mirror symmetrical structural units co-existed there. Partial dislocations having Burgers vectors which were close to half of a displacement shift complete lattice were introduced so as to connect a glide-mirror symmetrical structural unit with a modified mirror-symmetrical one in the grain boundary. The Burgers vectors were close to $a/18[122]_L\|$ $a/18[122]_R$ and $a/36[\bar{4}11]_L\|a/36[\bar{4}11]_R$.

The interactions of dissociated lattice dislocations with a near-$\Sigma = 3$ grain boundary in copper were analyzed[22], showing that Shockley partials were required in order to recombine when entering the grain boundary, so as to form an absorbed perfect lattice dislocation. Decomposition of the latter into 2 displacement shift complete products then occurred. Complex reactions between displacement shift complete dislocations led to further stress relaxation.

Alumina bi-crystals with low-angle and near-$\Sigma 3$ <00•1> tilt grain boundaries were prepared[23] by diffusion-bonding and the boundary energies were analyzed theoretically. Partial dislocations with Burgers vectors of 1/3<10•0> type were distributed periodically over the boundaries, together with stacking-faults. The length of the stacking-faults decreased with increasing misorientation angle. In the case of a near-$\Sigma 3$ grain boundary, an array of displacement shift complete dislocations with a Burgers vector of $1/3<1\bar{1}\bullet0>$ formed periodically along the boundaries. Those boundaries did not harbour stacking faults.

Relaxation of intergranular stresses linked to extrinsic grain-boundary dislocations was studied[24] in nickel grain boundaries of {311} and {332} $\Sigma = 11$ type. In strictly symmetrical periodic grain boundaries, incorporation of extrinsic grain-boundary dislocations in the form of displacement shift complete dislocations occurred.

Atomistic computer simulations were made[25] of $\Sigma = 5$ (310)[001] symmetrical tilt boundaries having a tilt-angle of 36.9°. Sliding was found to be controlled by grain-boundary dislocation activity with Burgers vectors which belonged to the displacement shift complete lattice.

A twin boundary in nm-sized gold contacts was sheared[26] within a high-resolution transmission electron microscope and the deformation was observed *in situ* at the atomic scale. The shearing produced sliding of the twin boundary, and simultaneous boundary migration. The relationship between sliding and migration agreed with the predictions of the coincidence-site lattice model and displacement shift complete theory.

When *in situ* straining experiments were performed[27] on nanocrystalline and ultra-fine grained aluminium at ambient and intermediate temperatures, both samples exhibited appreciable stress-assisted grain growth. The strain which was produced by grain-boundary motion in ultra-fine grained material was of the order of a few percent. The results could not be entirely explained in terms of displacement shift complete dislocation motion, and it was proposed that grain-boundary motion occurred via both shuffling and secondary displacement shift complete dislocation motion.

The stress-induced behavior of high-angle near-coincidence symmetric tilt boundaries was studied[28] in bi-crystalline zinc specimens with regard to the coupling between boundary sliding and migration. The angular deviation from the coincidence misorientation had a noticeable effect upon the sliding/migration ratio. Changes in the latter, with changes in boundary misorientation, were due to variations in extrinsic secondary grain-boundary dislocations. These results incidentally revealed the limitations of the coincidence site lattice and displacement shift complete lattice models for the quantitative description of the structure of near-coincidence boundaries.

A bicrystal of $SrTiO_3$ containing a $\Sigma = 9$, [110]/{221} symmetrical tilt grain boundary was experimentally observed and theoretically studied with regard to its most stable structure[29]. It was found that, when the grain boundary was slightly tilted away from the coincident site lattice orientation, displacement shift complete dislocations were introduced at the grain boundary, in order to accommodate the misorientation between the adjacent crystals, while the most stable structure remained unchanged.

The elastic and plastic deformation of copper bi-crystals with coherent and incoherent Σ = 3 ($\bar{1}11$) twin and Σ = 3 ($\bar{1}12$) grain boundaries was investigated[30]. The incoherent boundary suffered a reduction in elastic resistance due to an increase in excess free volume and structure-dependent modulus, whereas the coherent boundary had little effect upon elastic deformation. The propagation of plastic deformation was strongly blocked by dissociation into a displacement shift complete lattice dislocation, while plastic deformation of the Σ = 3 ($\bar{1}12$) boundary revealed the occurrence of easy slip-transfer across the interface.

The shear response of Σ = 3 [$\bar{1}10$]-tilt ($\bar{1}\bar{1}5$)/(111) and Σ = 9 [$\bar{1}10$]-tilt (115)/(111) asymmetrical tilt grain boundaries in copper and aluminium was studied[31], showing that the grain-boundary structure could be well-described by using coincidence site lattice theory. Room-temperature shear of these boundaries along 8 directions within the boundary plane showed that these boundaries can transform from one to the other by forming a coherent twin boundary. The transformation could take the form of sliding, coupled sliding-migration, faceting, etc, depending upon the shear direction and the material. Details of the transformation mechanisms were explained in terms of displacement shift complete theory. It was noted that, although the latter theory could be applied to the sliding-migration motion of grain boundaries, effects such as the shear direction within the boundary plane and the bonding characteristics of material could play an important role in the shear response.

Preparation of a MgO bi-crystal, containing Σ = 5 and near-Σ = 5 grain boundaries with a tunable bonding angle, demonstrated[32] that misalignments in grain boundaries could be compensated by a periodic displacement shift complete dislocation-array.

Shear-stress driven grain-boundary migration is a common phenomenon in small-grained polycrystalline materials. It was shown[33] that a displacement shift complete dislocation mechanism, for grain-boundary shear coupled with migration, still operated even when the orientation of the boundary deviated by a few degrees from the appropriate coincidence site lattice boundary. This implied that any large-angle grain boundary might be one for which shear-coupled boundary migration motion could occur via the displacement shift complete mechanism. It was concluded that the latter bi-crystallographic lattice structure was the main reason why grain boundaries could migrate under a shear stress.

A twist in bi-layer or few-layer graphene breaks local symmetry and introduces features such as new band-gaps. It was possible to determine[34] the atomic structure of such a twisted graphene in terms of a moiré superstructure, which was parameterized by a single twist-angle and lattice constant. Coincidence-site and displacement shift complete theories showed that the in-plane translation state between layers was not an important structural parameter, and that the current model was adequate for describing bi-layer and few-layer twisted graphene.

This is perhaps as good a place as any to make a plea to researchers *not* to capitalize routinely the term, moiré. There *was* never a 'Monsieur Moiré' to be made worthy of such eponymity.

It was noted[35] that the Brandon criterion is often used to quantify the distribution of coincident-site lattice boundaries in studies of grain-boundary character. This criterion is supposed to define the range within which special boundaries exist, but it has exceptions. Its limits were deduced in terms of deviations of the misorientation axis and of the boundary plane from the symmetrical tilt orientation, showing that deviation of the misorientation angle is more important than boundary-plane deviation. The limiting deviation for various orders of twin boundary is approximately constant at 1°. This result for various boundaries could also be obtained by assuming that secondary dislocations were spaced in proportion to the displacement shift complete lattice vector.

An equivalence of displacement shift complete lattice dislocations and grain-boundary kinks in graphene can be demonstrated topologically and energetically[36]. In the former case, a grain-boundary kink and a displacement shift complete lattice dislocation both translate the coincident site lattice. The energetic equivalence is meanwhile established by comparing atomistic and continuum elasticity models of metastable states, and showing that displacement shift complete dislocations are well described by elasticity theory. Continuum results can moreover be fitted to atomistic results by using a single adjustable parameter: the displacement shift complete dislocation core radius. Atomistic results show that low-Σ boundaries offer high energy barriers to grain-boundary motion, thus matching continuum results for smaller core-radius dislocations. Larger energy barriers for low-Σ boundaries are consistent with data on isolated low-Σ boundaries in graphene.

The above examples illustrate the usefulness of the displacement shift complete concept in a wide range of static and dynamic situations. In the present context however it is its application to heterostructural cases which is more relevant.

High-resolution electron microscopy of pure edge threading dislocations in GaN layers, grown by molecular-beam epitaxy onto (00•1) sapphire, showed[37] that the interface structures exhibited 5/7 or 8 atom cycles. These configurations were equally common for isolated dislocations and for low-angle boundaries. Examination of the coincidence grain boundaries showed that they were all made up of pure edge dislocations having the above atomic structures. The defects which were introduced by deviations from coincidence were associated with steps, and their Burgers vectors corresponded to the smallest vectors of the displacement shift complete set. Defect-free steps, which belonged to the sides or diagonal of the coincident site lattice unit-cell, occurred at the interfaces,. The reconstruction of some boundaries was possible only by taking into account the occurrence of structural units which exhibited 4-atom ring cycles for the dislocation cores. In non-symmetrical interfaces, a new structural unit made of 5/4/7 atom rings constituted the core of one grain-boundary dislocation.

When hexagonal γ-Nb_5Si_3 plates precipitate from niobium solid-solution in heat-treated Nb-Si-based alloys, the orientational relationships and interfacial structures are: $[111]Nb\|[00\bullet1]\gamma$ and $(10\bar{1})Nb\|(1\bar{1}\bullet0)\gamma$ with a near coherent interface. The interfacial structure has been explained[38] using coincidence site lattice and displacement shift complete lattice models. This predicted the formation of an array of phase-boundary dislocations at the interface. Those with a Burgers vector of $1/3<\bar{1}2\bar{1}>$, on a (111) niobium plane, were suggested to be able to promote the precipitation of γ-Nb_5Si_3.

An important type of heterostructural interface is that which occurs between the phases of an eutectic alloy. In directionally-solidified LaB_6-ZrB_2 composites, the orientational relationship between the two phases corresponded[39] to a high-symmetry near-coincidence site lattice, and the latter concept also predicted the predominant interface facet planes. A small (2 to 5°) tilt away from the high-symmetry orientation led to an increased volume density of coincident sites. The configurations of interfacial misfit dislocations were also in good agreement with the predictions of the displacement shift complete lattice and secondary original lattice models. The interfaces were then supposed to relax into relatively low-energy configurations.

The orientational relationship and interface structure between Ti_5Si_3 precipitates and γ-TiAl were investigated[40], showing that the habit plane of Ti_5Si_3 in TiAl was $(00\bullet1)\zeta\|(111)\gamma$. There were however no low-index parallel directions of the two phases in the plane. This anomaly was predicted by a geometrical method in which an overlap of the reciprocal lattice points of adjoining crystals was used to deduce the optimum orientational relationship. Although there was a marked difference in the crystal structures of TiAl and Ti_5Si_3, the interface was semi-coherent, well-matched and was associated with the largest possible displacement shift complete lattice that corresponded to the bicrystal.

The crystallography of S-phase precipitates was studied[41] in overaged Al-Cu-Mg alloys having a low Cu:Mg atomic ratio. The orthorhombic S-phase had the form of laths which were elongated, with $[100]S\|[100]\alpha$, in the face-centerd cubic solid-solution α-phase matrix. In a given sample, it could adopt a continuous or near-continuous range of orientational relationships which were characterized by rotations of up to $7°$ about common axes, defined by the limits: $(001)S\|(021)\alpha$, $[100]S\|[100]\alpha$ and $(0\bar{2}1)S\|(014)\alpha$, $[100]S\|[100]\alpha$. The variations in orientation were accommodated by changes in the interface orientation and structure and by a systematic variation in the lattice parameters of the S-phase. Such changes were, in turn, associated with changes in the α-phase lattice parameter; possibly involving defects.

A method which was based upon a combination of the Δg parallelism rule[42], coincidence site lattice and displacement shift complete lattice concepts was applied[43] to the habit-plane of δ-phase precipitates and linear defects in Inconel 718. A small scatter in the habit-plane orientations around an ideal rational plane was attributed to the existence of a mixture of 2 types of step, having differing heights and inclinations, that played an important role in maximising the degree of matching. They were associated with secondary dislocations having a Burgers vector of $1/6[11\bar{2}]\gamma\|1/3[001]\delta$, parallel to a near-invariant line along$[1\bar{1}0]\gamma$. The spacing of secondary dislocations, projected onto the terrace plane, was about 6.3nm.

0-Lattice Theory

Bollmann sought[44] matches by looking, assuming a given relative orientation of two crystals, for all of the locations where the two crystals matched. This was done by treating them as interpenetrating lattices. This then effectively defined 2 sets of positions for one set of atoms. A boundary which separated the crystals was then inserted so as to maximise the number of locations of best matching. Atoms were then imagined to be placed on one side of the boundary, in the positions of one crystal and, on the other side, into the positions of the other crystal. The concept of a three-dimensional construction of a best match could then be discarded, except in the boundary region. Instead of considering the atomic positions to be rigid, and the same as those in a perfect crystal, the boundary was relaxed in the sense that those atoms which were not in a best-match position were slightly displaced. At the same time, atoms in bad-match positions were contracted so as to form dislocations.

This provided a means for dividing those parameters, involved in determining the boundary structure, into two groups. One group was related to the crystal structures and their relative orientation, and the other was related to the path of the boundary. Each group could be handled independently. As is usual with situations involving materials of the same type, the method was largely applied to defects. A grain boundary was seen as consisting of a dislocation network with, between the dislocations, areas of perfect crystal which were slightly distorted elastically by the strain-field of the dislocations. The two crystals matched in those areas. It was noted that, in the coincidence site lattice approach, a coincidence site was an atomic site which was common to both lattices. The two crystals then matched in the sense that an atom occupied a lattice site in both crystals. It was envisaged that there existed a rule of so-called conservation of crystal structure, such that the tendency of a growing crystal to continue its structure extended into the boundary from both sides and helped to determine the boundary structure. In a grain boundary, deviations from the perfect crystal structure are compensated by dislocations. When the density of required dislocations became so high that little of the structure was left, the crystal could conserve at least a part of the original structure by forming a common superlattice: the coincidence site lattice.

That is, either the crystal structure is conserved, or a common superlattice is conserved. In the first case, a point of best-match can be defined as a coincidence of points which are in equivalent positions in both lattices; including here all points of the crystal entity and not just the atom positions. These points were termed 0-points and formed point-lattices within the interpenetrating lattices. The 0-lattice was the sum of all of the locations of best matching.

In calculating the 0-lattice, both crystals were first formulated as translation lattices in which only the translational periods of the crystals were of interest. A point-to-point correlation between the lattices was then defined by a linear homogenous transformation which could be a rotation by a small angle, an expansion or both. The transformation connected the closest neighbours of the two lattices in the neighbourhood of the origin, and so had to be a point-to-point correlation between the two lattices and to be the closest neighbour correlation (unless this led to a violation of volume conservation).

Solution of a rather difficult equation by using a clever trick led to an unique answer which permitted the 0-lattice to be constructed. The latter construction contained all of the information concerning the possible boundaries which could conceivably separate two crystals of given structure and relative orientation.

Once the 0-lattice cell structure is known, it can be used[45] to answer questions with regard to optimum boundaries[46]. When the 0-lattice consists of parallel 0-planes, a boundary which lies in an 0-plane is clearly more energetically favourable than one which is perpendicular to it. In the first case, the boundary does not intersect cell-walls and consequently does not contain dislocations, whereas it does in the second case.

When the structures of low-angle and high-angle grain boundaries were examined[47] using a dislocation approach, it was shown that coincidence-site lattice theory could account for the structures of boundaries of any angle. The methods used were quite general and could be applied to more complex cases. In particular, any grain boundary could be considered to arise from combinations of lattice dislocations which were associated with each of the grains comprising the boundary. Concepts such as the Burgers circuit were rationalized and the defect content which was described by a given circuit was shown to depend directly upon its reference lattice, whereas the strain fields were related to relaxations which occurred around the defect. It was consequently concluded[48] that the relatively complex 0-lattice theory might not be necessary to an understanding of the dislocation structures of grain boundaries. In response[49], it was suggested that this criticism was based upon an unrealistic model in that the dimensions of the atoms were not taken into account.

The theory was meanwhile applied[50] to martensitic transformations, where it was used to define the habit plane, and to the modelling[51] of ledges in partially coherent boundaries between body-centerd and face-centerd cubic materials. The method was here extended so as to permit the matching of all of the atom positions in the two crystal structures, as well as matching between relaxed atom positions. Misfit-compensating ledges and structural ledges were predicted to occur when interphase boundaries, which were made up of parallel steps and terraces, intersected additional 0-points. Misfit-compensating ledges with Burgers and Pitsch-Schrader terraces had Burgers vectors and inter-ledge spacings which were half of those associated with misfit dislocations on the corresponding flat interfaces. Bi-atomic structural ledges with Burgers and Potter terraces had Burgers vectors and inter-ledge spacings which were one-sixth of the misfit dislocation equivalents. In cubic materials, the dislocation structures[52] of grain boundaries were supposed[53] to change continuously as the misorientation changed according to 0-lattice theory, but they were later shown[54] to change discontinuously if grain-boundary phase transformations occurred with changes in composition or temperature. This was even more likely in hexagonal lattices due to the usually irrational c/a ratio.

A reciprocity relationship for the 0-lattice was later introduced[55] which completed existing reciprocity results for coincidence sites and displacement shift complete lattices. This indicated that it was possible to observe 0-lattices by using an electron microscope with an annular aperture and thus permit the study of strain-fields in an interface or between a thin growing film and a crystalline substrate.

Defect singularities are defined by the elimination of types of defect which must be present in vicinal interfaces. Singularities in dislocation and ledge structures have lately been integrated into the study of orientation relationships by means of 0-lattice theory[56]. Interfaces which exhibit singularities in the dislocation structure have been identified. An interface which is singular with respect to the interface orientation must be normal to at least one vector which connects 2 reciprocal points from different lattices, or must obey certain parallelism rules. Computer simulation methods were also used[57] to determine the dislocation network at the interface of cubic crystals by means of 0-lattice theory, using firstly a single 0-cell and then multiple 0-cells.

A geometrical technique, based upon 0-lattice theory, was developed[58] in order to permit the approximate atomic simulation of crystal boundaries, including a transition zone in which atoms were placed between sites of the lattices on either side of the boundary. Atoms could also be removed when they were separated by less than some two-thirds of the equilibrium spacing. The model was applied to interfaces between face-centerd and body-centerd cubic crystals, and these were interpreted in terms of matching and non-matching areas which, contained so-called coherency, and misfit, dislocations. The latter occurred where the ledge structures of the two crystal surfaces lost registry, while the former were found where they corresponded.

A further problem with the 0-lattice theory is that, in systems having an invariant line or an invariant plane, the required 0-elements may not exist in three-dimensional space[59]; thus limiting its applicability to the relevant interfaces[60]. In order to treat dislocation structures in interfaces for which the 0-elements do not exist in three-dimensional space, a generalized 0-element approach has very recently been introduced which exploits the Moore-Penrose pseudo-inverse. The generalized 0-elements, seen as being least-squares solutions to the 0-lattice, then play a role which parallels that of the ideal 0-elements and extends the list of candidate locations for coherent regions between dislocations. The predicted interface dislocation structures in both homo- and heterostructural systems are found to be in good agreement with experimental observations and with molecular statics and molecular dynamics simulations.

Simple geometrical approaches to interface crystallography, based upon invariant-plane and invariant-line deformation criteria, were reviewed[61] from the point of view of matrix algebra and a re-formulation of phenomenological crystallographic theories of martensitic transformations using the infinitesimal-deformation approach. As an application of the invariant-line deformation criterion, epitaxial relationships between thin-film crystal deposits and substrate crystals were analyzed. The simple geometrical criteria were found to be very useful in explaining features of interface crystallography. Experimental and theoretical studies of epitaxial relationships between face-centerd cubic and body-centerd cubic crystals, either as deposits or substrates, were reviewed[62] and simple criteria based upon geometrical lattice-matching and misfit strain-energy considerations were used to predict favourable epitaxial relationships. Excellent agreement with the criteria was found; not only for low-index substrate planes but also for higher-index, and curved, substrate surfaces.

As an example of a very practical application of 0-lattice theory to the design, rather than explanation, of interfaces there is the case of a tailored interface which can lessen the effect of helium-induced damage by encouraging the precipitation of the gas within continuous linear channels[63].

Another related geometrical method, proposed for the calculation of interfacial dislocation structures, assumes a periodic correspondence between the structure within a good matching site cluster and a conserved structure between dislocations[64]. Each interfacial dislocation is associated with a couple of correlated Burgers vectors, and a set of vectors from each real lattice is identified according to the translational symmetry within a good matching site cluster. The 0-lattice theory is then used to quantify the distribution of good matching site clusters at the 0-elements, and poor matching regions at the 0-cell walls. The configuration of dislocations in a general semicoherent interface is finally predicted according to the effective 0-cell wall traces in the interface.

A recurrence relationship can be shown to exist between the 0-lattices of rotation-related grain boundaries when a suitable parameterization of the rotation-angle is introduced[65]. This then permits the basis vectors of any 0-lattice to be found by simple vector-addition when the basis vectors of any 2 orientations are known. It also serves to separate the angular space into disjoint sets, and hence splits grain boundaries into a number of classes, regardless of the crystal system.

Periodic good-matching bands, centered on 0-lines, can be seen as being a characteristic feature of the structures in irrational singular interfaces[66]. A quantitative analysis of the distribution of good matching zones, with regard to plane-matching geometry, makes it clear that the matching of one set of principal planes does not constitute good lattice matching. The latter is possible only at locations where 3 sets of non-linearly related moiré planes intersect. The matching of at least one set of principal planes in an interface implies the presence of periodic good matching zones in the interface. The distribution of exact intersections could be deduced on the basis of 0-lattice theory, and the approximate intersections could be used to predict a possible interfacial structure when periodic 0-elements do not exist.

The validity of the 0-lattice theory for predicting grain-boundary structure has been checked[67] by comparing the predictions with high-resolution transmission electron microscopic images. This showed that the predicted periodic structure of a given <110> symmetrical tilt grain boundary was consistent with those observed, regardless of the nature of the atomic bonding. There was a good correlation between the misorientation dependence upon grain-boundary dislocation density and that upon grain-boundary energy. This suggested that most of the grain-boundary energy arose from the elastic and core energies of grain-boundary dislocations. The energy per grain-boundary dislocation was less dependent upon the magnitude of the Burgers vector, but depended more upon the Burgers vector divided by the grain-boundary dislocation density; a quantity which lay within the range of 0.1 to 1.

The macroscopic and microscopic crystallographic parameters which can be used to characterize fully the misorientation across a grain boundary were defined[68] early on. The definitions were then used to identify coincidence sites along a line, over a plane or within a spatial sub-lattice. Such a coincidence was defined in particular for twin boundaries in a cubic crystal, and was related to 0-lattice theory. General techniques for the calculation of grain-boundary structures were described, but the limitations of methods which are based upon purely geometrical considerations were nevertheless pointed out.

In many cases the structure of an interface can be explained[69] in terms of so-called good-matching islands which are separated by misfit dislocations. The 0-lattice theory can predict the dislocation-spacing, but not the core-width. A so-called localization-parameter permits the prediction of the width of misfit dislocations, where the parameter is equal to the bonding-energy of the two crystals, divided by their average cohesive energy. A survey of the localization parameters for some 30 metal/metal, metal/ceramic and ceramics/ceramic interfaces suggested that the misfit dislocations were delocalized when the parameter was below 0.25.

A recently developed program, the 'Arrangement of Interface Dislocation Arrays', identifies[70] all of the dislocation networks which satisfy the quantized Frank-Bilby equation for any interface between two single-atom cubic crystals. It reveals the possible range of dislocation patterns at an interface and can be used to analyse experimentally observed interface defect structures.

With regard to specific materials, an early application[71] of 0-lattice theory was to electron microscopic observations of high-angle tilt boundaries in stainless steel bi-crystals with a [001] tilt axis. Deviations from coincidence orientations appeared as pseudo-subgrain boundaries, and the low stacking-fault energy of the stainless steel had a marked effect upon the structure of the boundary. A three-dimensional invariant-line strain model[72] for face-centerd to body-centerd cubic phase transformations was used to predict the growth direction and habit plane of lath-shaped precipitates. The results were almost identical to those deduced using 0-lattice theory and a Kurdjumov-Sachs relationship. Both theories agreed well with experimental observations of chromium-rich body-centerd cubic precipitates in Ni-45wt%Cr alloy. The growth direction of the precipitates was some 512° away from the common close-packed directions of the two lattices along an invariant line. The observed habit plane, {112}, was the unrotated plane of the phase transformation. A second facet plane of the precipitates, close to {313}, was also predicted by using 0-lattice theory. The structure of a low-angle grain boundary in zinc was examined[73] in order to check the applicability of 0-lattice theory to hexagonal closed-packed crystals. The tilt boundary contained a dislocation array with a Burgers vector of [00•1] (0.495nm), and exhibited an unexpected step configuration. The Burgers vector of the principal dislocation array, the step dislocation, the dislocation spacing, and the average boundary plane of the boundary all strongly supported 0-lattice theory, but it failed to predict the structure of the boundary.

High-angle tilt boundaries (110)∥(001) in silicon were created[74] by direct wafer bonding. Reconstruction of the interface along the [$\bar{1}$10]∥(001) direction was carried out within the framework of 0-lattice theory demonstrating that, in order to preserve covalent bonding across the interface, it should comprise {$\bar{1}$11}∥{$\bar{1}$12} facets which are intersected by up to six {1$\bar{1}$1} planes with three 90° Shockley dislocations per facet. The formation of Frank dislocations was predicted on sufficiently long {$\bar{1}$11}∥{$\bar{1}$12} interfaces, with a period which was equal to 6 times that of Shockley dislocations. Long-range undulations of the interface are attributed to a deviation from an exact 90° tilt of the layer with respect to the substrate. This had an effect upon the Burgers vectors of dislocations which adjusted the in-plane twist misorientation[75].

The structures of low-angle tilt boundaries in NiO were studied[76] by means of electron microscopy, showing that both boundaries were faceted; with one having a complex dislocation content and facets with a period of about 8.8nm. When 0-lattice theory was used to predict the geometry and dislocation-content of the facets, the predictions were in generally excellent agreement with the observations. It was suggested[77] that a tilt boundary having an arbitrary rotation axis would be faceted. A procedure was proposed[78], for analysing the crystallography of arbitrary grain boundaries, which was based upon 0-lattice theory. The configurations of boundary dislocations which were associated with those Burgers vectors were then refined by taking account of dislocation reactions. After applying the procedure to grain boundaries in orthorhombic $YBa_2Cu_3O_{7-\delta}$, it was concluded that the methodology was suitable for determining the structure of arbitrary grain boundaries in low-symmetry crystals.

Coming finally to a heterostructural example, 0-lattice theory was applied[79] to {222}MgO/Cu-Ag interfaces which had been prepared via internal oxidation. The observed spacing (1.45nm) between misfit dislocations on a <110> projection agreed with 0-lattice predictions and it was concluded that these interfaces were semi-coherent and contained a trigonal network of pure edge misfit dislocations lying parallel to <110>-type directions, with an (a/6)<211>-type Burgers vector. Misfit dislocations were also found in a stand-off position, at a distance which was equal to a single (111) spacing of the Cu-Ag matrix.

Among the most widely studied interfaces in cubic materials is the $\Sigma = 1$ interface, which includes all low-angle grain boundaries in homogeneous materials and any phase boundaries in which the corresponding planes and directions of the adjoining materials are essentially parallel. Misorientations and misfits are then compensated for by the creation of dislocations in the material or at the interface. Another type of interface is that of the $\Sigma = 3$, first-order twin, boundary. This is parallel to common planes in two crystals, as between (111) and (112) facets in metals, or is parallel to different planes in two materials. A third type of interface is the (111)/(100) interface of the cubic system. It is found both between materials having the same composition and structure, and between materials where the compositions, structures or lattice parameters differ. The interface can also contain facets which involve a different relationship. This orientational relationship occurs in bulk and thin-film metallic, oxide and semiconductor systems and the pseudo six-fold and four-fold symmetries, respectively, of the (111) and (100) planes are already very different and pose interesting matching problems.

The stability of the interface has been surveyed[80] by using models for predicting the interfacial energy, and the interface is thus a benchmark for predicting more general cases of solid/solid interface stability. The mechanisms and models which have been used to treat this interface are very diverse, but the fact, that abutting low-index planes are parallel there, appears to be the main common factor favouring this interface across a range of systems. Because the preference for this interface does not depend upon the interatomic bonding of a material, it is deduced that short-range interactions govern its stability.

Application of the coincidence site lattice theory, originally intended for treating homogeneous interfaces, suggests that the density of common sites which are shared by the two lattices should be a determining factor in deciding whether a particular interface can exist. The volume-density of common lattice sites is given by $1/\Sigma$ and, because grain boundaries can have low Σ-values, the latter tends to be associated with a low interfacial energy. For a cubic material, the lowest Σ-value corresponds to a first-order twin-boundary in a face-centered cubic lattice. On the other hand, this is the so-called bulk Σ-value and it seems more reasonable to assume that the density of coincident sites within the interface plane is more relevant. The presence of secondary dislocations in low-Σ boundaries nevertheless indicates that grain boundaries can lower their overall energy by locally adopting the exact Σ structure.

A so-called lock-in model has been proposed for low-energy interfaces in which close-packed rows in the surface of a metal lock-in to the valleys in the surface of an underlying ionic crystal. The correlation between the interfacial energy and the density of lock-in configurations is expected to depend upon the atomic-radius difference between the metal and the ionic material. Entropy considerations may be may be important here because the interface bonding is very different to that of either bulk material. The lock-in model can also be viewed as being a form of epitaxy at the atomic scale, because the surface topology governs the orientation of the growing phase.

In another view of interphase surfaces, it has long been known that internal interfaces and external surfaces both exhibit a marked tendency to have facet-planes which are parallel to low-index planes. This can be generalized to the idea that low-energy interfaces are governed by the requirement that the planes on each side of the interface should be densely-packed ones. A low Σ-value may thus be a necessary but not sufficient condition for the existence of a low-energy interface: the area of the unit coincidence site lattice cell on the boundary plane, and the interplanar spacing of the lattice planes parallel to the interface plane, may both be important criteria for the establishment of low-energy interfaces.

The ubiquity of (111)/(100) phase boundaries is quite striking. In directionally-solidified NiO/ZrO_2 eutectic, the interface plane is parallel to $\{111\}NiO\|\{100\}ZrO_2$ and the growth direction is parallel to $<1\bar{1}0>NiO\|<001>ZrO_2$. It is perhaps significant that the interface plane is a common oxygen-atom plane, while the $\{111\}NiO$ and $\{100\}ZrO_2$ planes contain alternating layers of cations and anions. The considerable difference in the lattice parameters of the oxides also means that alternating oxygen atoms on the $\{100\}$ plane of ZrO_2 almost coincide with one on the $\{111\}$ plane of NiO.

Upon applying the embedded-atom method to various twist boundaries on the $\{100\}$. $\{110\}$ and $\{111\}$ planes of silver and nickel, several low-energy interfaces are predicted to exist, including the $\{111\}Ag\|\{100\}Ni$ interface, with a common $\{01\bar{1}\}$ direction and an interfacial energy of 470mJ/m^2. In the case of the $\{111\}Ni\|\{100\}Ag$ interface, the interfacial energy was 670mJ/m^2. The difference was attributed to an anisotropy in the contribution which chemical energy made to the interfacial energy. In experiments, the deposition of silver onto $\{111\}Ni$ first results in the appearance of $\{111\}$-oriented silver islands. These are then engulfed by $\{100\}$-oriented material, leaving a continuous silver film in spite of this having a theoretically higher energy. The initial nucleation of low-energy epitaxial islands seems to be the driving mechanism in the early stages of growth, before kinetic factors start to determine the final orientational relationship. Lattice-matching alone clearly cannot explain all of the experimental observations.

In metal/oxide systems are found orientational relationships such as [010]MgO∥[11$\bar{2}$]Cu; (100)MgO∥(111)Cu. In internally-oxidized Cu-Mn alloy, there occur interfaces in which the (111) planes of copper match (100) planes of MnO by way of a 55° rotation about their common [01$\bar{1}$] direction. Another common orientational relationship is a 71° rotation about <011>. These observations indicate the existence of crystallographic relationships which are based upon common close-packed directions and parallel close-packed planes. Theories which are based upon stable orientational relationships must however take account of chemical anisotropies at the interface. The growth of epitaxial aluminium films on {111}Si results in a microstructure of <100>-oriented aluminium rotated by 60° around <110>, which is a common direction in both the substrate and the film. The growth of aluminium films on {100}Si however rarely results in epitaxy. It is surprising that, for example, monocrystalline (111) aluminium films can be grown on Si(111) at room temperature, given the 25% lattice- mismatch[81].

Growth of layers of ErAs on {100}GaAs, followed by a GaAs overlayer, produces regions of {111}-oriented GaAs overlayer on {100}ErAs, showing that lattice-mismatch does not alone control the epitaxial alignment of interfaces. The mismatch of {100}GaAs on {100}ErAs is only about 1.6%, but {111}GaAs∥{100} has been observed; with the interfaces being abrupt and clean. A possible explanation of these and other observations has been suggested to be the relative surface energies of ErAs and GaAs: the nucleation of {111} GaAs may occur due to the presence of sub-monolayer impurities in the ErAs. During the first few monolayers of the growth of GaAs, gallium supersaturation in the islands may make <111> the preferred growth direction.

Epitaxial interfaces in semiconductor/semiconductor systems might be suspected to be quite simple because the directionality of the bonds in each material can be preserved at the interface, but both {100}/{100} and {111}/{100} orientational relationships have been observed in the case of CdTe/GaAs. Two common epitaxial interfaces are {111}CdTe/{100}GaAs and {100}CdTe/{100}GaAs. Differing growth methods reliably lead to the creation of {111}CdTe films on {100}GaAs, with a 14.6% lattice-mismatch along the common <110> direction which is parallel to the growth interface. An 0.7% mismatch between {11$\bar{2}$}CdTe and {110}GaAs planes is expected to affect the epitaxy. It is proposed that an arsenic-depleted GaAs surface induces {111} epitaxy whereas a desorbed oxide layer on GaAs promotes {100} epitaxy. Micro-twins and stacking-faults are common in {111}CdTe microstructures because two equivalent <111> orientations are possible, via a 180° rotation about <100>GaAs. An understanding of the detailed interfacial chemistry, as well as the use of lattice-matching arguments, is clearly essential when predicting epitaxial behaviour on compound semiconductors. Highly-oriented {111}-oriented films of MgO have been produced on {100}GaAs by using pulsed-laser ablation, with the epitaxy being perhaps promoted by traces of native GaAs oxide on the surface before deposition. The observed epitaxy between MgO and GaAs had certainly been influenced by the crystallography of the substrate. The growth of $\{111\}$CdTe film, by evaporation onto freshly-cleaved $\{001\}$NaCl having a common $<1\bar{1}0>$ direction, has been explained in terms of the lock-in model.

Grain boundaries having large Σ-values have been studied in sintered polycrystalline $MgAl_2O_4$ and, in spite of large misorientations, they exhibit faceting at the micron scale, so that low-index planes of both grains can lie almost parallel to each other. The $\Sigma = 41$ and $\Sigma = 99$ boundaries have a common $<110>$ direction, about which the grains are rotated. On the basis of coincidence site lattice theory, the $\Sigma = 41$ grain boundary is not predicted to be so favoured. Atomic-relaxation studies of high-Σ boundaries in face-centerd cubic metals suggest that the atoms try to maintain their bulk coordination. This then causes low-index planes to lie parallel to the interface, even when the interface is essentially incommensurate.

Like the $\{111\}/\{100\}$ interface in cubic systems, there exist non-cubic systems where interfaces form between two surfaces of differing crystallographic symmetry. In the case of the hexagonal lattice, the $(00\bullet1)$-plane has essentially the same symmetry as that of the $\{111\}$ planes in a face-centerd cubic lattice. Thus the growth of ZrO_2 films on $(00\bullet1)Al_2O_3$ produces interfaces in which the $\{001\}$ planes of the zirconia are parallel to the basal planes of the alumina, with the interface resembling the $\{111\}/\{100\}$ interface in cubic materials. That is, a pseudo-hexagonal surface is in contact with a surface having four-fold symmetry. In some non-cubic oxide systems, the interfaces often comprise crystal surfaces which are low-index planes. The $\{111\}/\{11\bullet0\}$ and $\{112\}/\{00\bullet1\}$ spinel/alumina interfaces are typical cases in which the symmetries of the planes forming the interface are different. Elsewhere, the $\{111\}$ planes of the spinel can be parallel to the $\{11\bullet0\}$, $(00\bullet1)$ or $\{11\bullet3\}$ planes of the alumina. One detail to be borne in mind here is that the spinel does not distinguish between the $\{11\bullet3\}$ and $\{11\bullet\bar{3}\}$ planes, or between the $(11\bullet0)$ and $(2\bar{1}\bullet0)$ planes, of the alumina. Because these planes are, in alumina, non-equivalent simply on the basis of the cation sub-lattice, it is to be concluded that oxygen ions play an important role in controlling the orientational relationships.

Epitaxial indium zinc oxide ($In_2Zn_2O_5$) thin films can be obtained[82] on $<100>$-oriented silicon, GaAs and InP by means of pulsed laser deposition. The crystalline quality of these thin films is similar to that of bulk single crystals. Increasing the deposition temperature causes the thin films to have a fine nanostructural form. Good lattice-matching between the highly-oriented $In_2Zn_2O_5$ layers and the substrate is possible without needing any buffer layer.

In one of the earliest systematic attempts[83] to model heterostructural interfaces, it was pointed out that there was an increasing number of important practical cases, such as (111)CdTe on (100)GaAs, where the two materials had entirely different crystal structures and where the previous key criterion of a lattice-match no longer applied.

A new criterion was proposed, which was to compare the interface translational symmetry with that of the materials on each side of the interface, instead of comparing the lattice-parameters. Two lattices were then said to match if the interface translational symmetry was compatible with the symmetry on each side of the interface. As an example, two-dimensional translations parallel to the $(11 \bullet 2)$ face of Al_2O_3 and the (111) face of silicon were considered. Superlattices were then formed by using multiples of the unit cells, such that 21 sapphire unit cells and 40 silicon unit cells could be equated: the sapphire superlattice cell being 15.386Å by 33.315Å and the silicon superlattice cell being 15.361Å by 33.258Å; the mismatch in each direction amounting to just 0.2%. The Al_2O_3 $[\bar{1}1 \bullet 1]$ direction was then parallel to one of the 3 equivalent $[01\bar{1}]$, $[10\bar{1}]$ and $[1\bar{1}0]$ silicon directions.

This arrangement was not claimed to be the true growth form of (111)Si on sapphire (101), but merely a demonstration of the existence of interface orientations which match to high precision. It is clear that this method was essentially a transfer, of the coincidence-site lattice concept, from high-angle grain boundaries in a single material to the hetero-epitaxial case. The presence of a finite mismatch unfortunately imposes the abandonment of the geometry-of-numbers methods which are so useful in coincidence-site lattice theory.

It was stressed that the local chemistry of the interface was not considered. The lattice match was characterized by just two parameters: the closeness of the match and a minimum unit-cell area. The above unit-cell area was $511Å^2$, and this was the smallest one which permitted a mismatch of less than 1% to exist in the unit-cell sides and angle. This implied a necessary compromise between match-precision and the superlattice cell-size: consideration of very large superlattice cells increased the number of suitable cells and the probability of a precise match. A larger interface unit-cell size unfortunately made it less likely that other forces could sustain the lattice-match. The match-precision placed a lower bound on the changes in lateral interatomic distances, in that a 1% mismatch of one of the sides of the common unit-cell would have to be accommodated by an average ±0.5% lateral movement of each atom on both sides of the interface; with some distances changing by more than 0.5%.

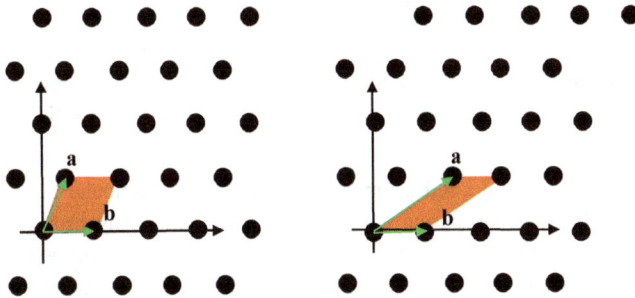

Figure 2. Different choices of base for the same lattice,
primitive (left) and non-primitive (right)

These changes would then produced strains and a thick epitaxial film, for example, would respond by creating defects. The common unit-cell area therefore had to be limited to a value which was compatible with interfacial periodic reconstruction. It was recalled that the free (111) face of silicon can spontaneously reconstruct to form a (7 x 7) unit cell having an area of $625Å^2$, thus making reasonable the existence of unit-cells having an area of some hundreds of $Å^2$. It was shown that a limit of 1% on match-precision and a limit of $600Å^2$ on unit-cell area could explain experimental results for (111)CdTe/(100). In general, this reduced the problem to a computer-search, over pairs of materials and over various directions, for potentially good matches.

When thin aluminium layers were prepared[84] by Knudsen-type evaporation onto {111} silicon surfaces, epitaxial growth was found on clean substrate surfaces with a (7 x 7) reconstruction with the epitaxial relationship: {111}Si||{001}Al; <110>Si||<110>Al. The epitaxial films had an overall twelve-fold symmetry axis, due to a combination of the 4-fold and 3-fold symmetry of the {001} deposit-plane and substrate surface, respectively.

The basic philosophy was that any slice through a three-dimensional lattice produced a surface having two-dimensional translational symmetry, and any reconstruction of atoms near to the interface generally reduced the symmetry. The symmetry-group of the reconstructed surface was therefore a sub-group of the symmetry-group of the unreconstructed surface. The lattice-matching task was thus reduced to scanning two-dimensional slices through 2 given lattices and comparing the resulting two-dimensional lattices in order to identify a common superlattice. This comparison encountered 2 difficulties: similar lattices might be rotated with respect to each other, or might be rotated mirror images of one another. In addition, there was in fact no simple way in which to characterize a lattice. That is, the lattice is uniquely specified by its primitive translations: these being the set of vectors whose linear combinations and integer coefficients position every lattice-point. But whereas the set of primitive translations defines a lattice, there is an infinite number of alternative sets of primitive translations which can generate a prescribed lattice (figure 2).

This could be done by using a reduction scheme, in which a particular set of primitive translations was chosen from among all of the possible ones. Only properties such as lengths and angles were used to choose the primitive translations, with no appeal being made to any particular coordinate system. No lattice rotation nor reflection, which would effectively constitute a different coordinate system, thus had no effect upon the choice of the primitive translations. A reduced set of primitive translations, **a,b**, was defined in which **a** was the shortest possible non-zero vector of the lattice, **b** was the shortest possible lattice vector that was linearly independent of **a** and the angle between **a** and **b** was non-obtuse.

The computer-search strategy then involved successively adding or subtracting the shorter translation vector from the longer one until no further reduction in size of the translation vectors was possible. The sign of one of the vectors was reversed if the angle between the two vectors ever became obtuse. This method also preserved the area of the unit cell. A lattice could have more than one set of primitive translations, but the lengths and angle were unique.

A further problem was to decide when 2 two-dimensional lattices in fact matched. The criterion used was that 2 lattices were judged to match if each one of them had a superlattice, and the 2 superlattices were the same: up to rotation, reflection and choice of primitive translations. The search turned out to be less daunting than anticipated because the unit-cell area of the superlattice was an integer multiple of the unit-cell area of the original lattice. For a lattice having a unit-cell area, A, it was necessary to consider all of the superlattices having a unit-cell area, nA, where n was an integer. There was only a finite number of such superlattices; certainly less than the sum of the divisors of n. A two-dimensional lattice for example would possess 4 or fewer distinct superlattices having a unit-cell that was 3 times larger than that of the original lattice, and a superlattice having a unit-cell which was 6 times larger than the original lattice would posses no more than 12 (= 1+2+3+6) distinct ones.

The method was applied to several practical examples. Both CdTe and GaAs have a face-centerd cubic zincblende structure with lattice parameters, a_0, of 6.481Å and 5.653Å, respectively. In both cases, lattice translations parallel to the (100) face form a grid of square ($a_0/\sqrt{2}$ x $a_0/\sqrt{2}$) unit-cells oriented parallel to the [01$\bar{1}$] and [0$\bar{1}$1] directions. The translations parallel to the (110) face form a grid of rectangular cells ($a_0/\sqrt{2}$ x a_0), with the shorter edge parallel to [1$\bar{1}$0] and the longer edge parallel to [001]. The lattice translations parallel to the (111) face form a grid of 60° rhombi ($a_0/\sqrt{2}$ long), parallel to [1$\bar{1}$0], [10$\bar{1}$] or [01$\bar{1}$].

Attention was then concentrated on the most important faces, (001), (110), (111), of GaAs and CdTe. This gave 9 possible combinations of areal ratios. For (100)/(100), (110)/(110) and (111)/(111), the unit-cell areal ratios were all equal to 1.314. This ratio could be approximated by various rational numbers, including 4/3, 13/10, 17/13 and 21/16; with errors of 1.4, 1.1, 0.5 and 0.1%, respectively. The ratio could clearly be approximated to any desired accuracy.

Choosing the approximation, 4/3, one then looks for a super-cell which is 3 times larger for CdTe and 4 times larger for GaAs. From these unit cells, superlattices are formed by juxtaposing copies of the same cell. It was noted that one of the [001] directions perpendicular to the (100)GaAs face could align itself with one of the [11$\bar{2}$] directions perpendicular to the (111) face of CdTe. The other GaAs [011] direction could align itself with CdTe [1$\bar{1}$0]. Any match of CdTe(111)/GaAs(100) having a different orientation had to involve a superlattice cell which was larger than 400Å^2, but this had been ruled out as part of the search procedure.

Also checked were all of the possible ways of matching one of the (100), (110), (111), (210), (211) and (221) faces of CdTe to one of the (101), (0$\bar{1}$1) and (111) planes of sapphire, the structure of the latter being rhombohedral with $a_0 = 5.1286$Å and $\alpha = 55°$ 17.36'. The (101), (0$\bar{1}$1) and (111) sapphire faces corresponding to the (1$\bar{1}$•2), (1$\bar{2}$•0) and (00•1) faces of hexagonal notation. Among the 18 possible face combinations, it was found that only 10 matched within the limits of a maximum 1% error and a superlattice cell area of less than 600Å^2. In some experimental work[85], a CdTe epitaxial film was oriented with its (111) plane parallel to a sapphire substrate surface. This was considered to be remarkable because it occurred in spite of the large (circa 4%) lattice-mismatch between CdTe and sapphire, and because of the absence of any threefold symmetry of the (1$\bar{1}$•2) and (1$\bar{2}$•0) sapphire substrates.

A more recently developed algorithm[86] rapidly scans the possible surface configurations of two monocrystals and identifies those pairs which may form stable hetero-interfaces. As in the original method above, the 2 crystals are sliced using various planes and all of the possible heterocrystalline interfaces are generated, based upon geometrical criteria and the predicted bond-directions of the atoms on each of the matching surfaces. Each configuration is assigned two ratings which are based upon deviations of the interfacial bond-lengths from the ideal value and upon the electronegativity differences between the atoms on each side of the interface. Considering the scheme now in more detail, the initial bond directions of atoms on the surfaces were used to set up the relative configuration of the planes. For each atom located in the bulk, those vectors pointing in the direction of its nearest-neighbours were identified; the nearest neighbors being found by calculating the distances between each atom, and all of the other atoms, and comparing them to the expected bond distances. The latter was in turn defined as being the sum of their atomic radii. Broken bonds at the 2 surfaces were found by taking scalar product, of a bond-vector in the bulk and the unit vector normal to the plane. A positive scalar product was thus associated with a broken bond. Assuming that a given atom on one of the surfaces was likely to form new bonds pointing in the direction of a broken bond, the scalar products of all broken bond vectors - one from each plane – were considered. When the scalar product was equal to -1, the broken-bond vectors pointed in exactly contrary directions and a bond was expected to form in the direction of the vectors and with the atoms separated by the summed atomic radii.

When considering the scalar products of broken-bond vectors which corresponded to atoms on the 2 surfaces, only atoms in the superlattice unit cells were included. For each scalar product which was within a certain tolerance, the surfaces were translated so that corresponding pairs of atoms lay at the origin and were shifted in the direction of an intermediate bond vector; determined from the difference between the 2 matching broken-bond vectors. The length was equal to the ideal distance between atoms.

Each initial configuration involved placing a super-cell of one material above an extended surface which consisted of 3 x 3 super-cells of the other material, thereby guaranteeing that all of the interactions between the atoms of both planes were accounted for.

The various planes and configurations were then given scores which were based upon nearest-neighbour distances and electronegativity differences between the atoms of the 2 surfaces. The former score ensured that the surface atoms were not linked by bonds which were too different to the ideal bond-distance. The other score checked that the electronegativity differences of atoms at the interface were maximized. These scores thus ranked configurations based upon their chemical compatibility. Density functional theory could be used to rank the various configurations more quantitatively.

The nearest-neighbour distance criterion was based upon evaluating the distance between atoms in the super-cell and those on the opposing surface after the initial configuration was established. The main aim was to calculate deviations from a supposedly ideal distance. The squared deviations, taken over all of the atoms inside the super-cell, were

Materials Research Forum LLC

https://doi.org/10.21741/9781644900475

averaged for each surface. The final score for each configuration and plane-combination was found by averaging the two scores, with values very to zero being sought as this signalled a good configuration.

The electronegativity-difference criterion for each atom within a super-cell on a given surface was then applied. The electronegativity differences between atoms were based upon the Pauling-scale. The average electronegativity differences for atoms inside a super-cell on each surface were calculated, and averaged over all of the atoms. The final score was divided by the maximum electronegativity difference between 2 atoms on each surface. The best type of configuration was expected to maximize the average electronegativity difference between atoms on the surfaces, and so scores close to unity were sought. In the case of single-element crystals, scores of unity were found when all of the atoms on the two surfaces had at least one nearest neighbour. As in the earlier research, the GaAs/CdTe case was again used to test the present lattice-matching algorithm.

Another recent innovation[87] has been to automate the high-rate screening of interfacial systems. The program permits the high-rate handling of nanoparticle/ligand combinations by taking account of the crystallographic planes of surface facets, ligand binding-sites on each facet and the ligand surface coverage for a given facet and ligand-binding site combination. This is obviously a much more general approach than those described above, and so only those applications which parallel the previous ones will be picked out here.

In general, the interfaces are modelled by using slab models of specified thickness and crystallographic structure, with ligands adsorbed on their surfaces. The choice of the slab model used to describe an interface requires specification of the bulk phase making up the given surface, the crystallographic facet of interest, the initial binding site at the surface, the atom on the ligand that is expected to be the nearest neighbour to the chosen binding-site on the surface, the approximate initial separation between the surface binding site and the atom and the ligand coverage per unit area of the surface.

Ligands are then placed above a slab that is generated from a bulk structure by specifying the Miller indices of the required facet. This is rather like the crystal-slicing stage of the above approaches. The program can be generalized so as to handle arbitrary facets having molecular species, adsorbed on the surface at any site, and then search for the most stable ligand binding configuration. The initial ligand configuration is defined by the position and orientation of the ligand relative to a reference surface atom on the surface, and the ligand can be rotated as required.

One search strategy would be to place the ligand randomly at a position above the crystal facet and randomly rotate it. Relaxing the resultant configurations then explores the local minima among them. This approach has been applied to lead acetate ligands on the (100)PbS surface.

In the case of nanocrystal surfaces capped with ligands, the surface energy is a function of the ligand coverage. Given the magnitudes of those energies for all of the facets of interest, it is possible to predict the multifaceted equilibrium shape which minimizes the surface energy. As is well known from classical crystal theory, this shape can be deduced by using the Wulff construction, and the present method extends its application, to cubic symmetry, to any space-group.

Returning now to heterostructural interfaces, the first step in the study of junctions between solid-state materials is, as usual, to construct the interface between the two crystal structures. Again as usual, there is the problem that there generally exist numerous possible coincident site lattices. An additional complication is reconstruction, and the multiple possible positions of atoms at the interface. Here again, brute-force approaches can identify those coincidence site lattice vectors between two surfaces which give the smallest lattice mismatch. The present program, using the original technique above, explores various interface configurations of the interface between two crystals and identifies the pairs which offer a specified lattice-mismatch, symmetry-matching and uniqueness. Many surfaces are known however to exhibit complex surface reconstructions which depend upon factors such as the temperature and partial gas pressure. The present algorithm permits the insertion of complex reconstructed surfaces before proceeding to heterostructure generation.

The methods so far described obviously have relevance with regard to epitaxy. Caution must however be exercised because a monolayer, perhaps not even a complete one, will have many more degrees of freedom than does an adjacent atomic layer which is part of another bulk material. This is found to be particularly true in the case of epitaxial organic layers. For example, the development of organic semiconductors during recent years – especially in thin-film form - have led to great interest being taken in the structure and properties of epitaxial molecular films.

The Curie symmetry-law has been applied[88] to the two-dimensional point symmetries of interfaces. Pseudomorphism and misfit dislocations can accommodate epitaxial matching problems if the layer and substrate both have the same crystal structure and the difference between the lattice constants is small. The layer then tends to maintain the crystallographic orientation of the substrate, while any small geometrical mismatch is handled by the appearance of misfit dislocations and elastic strain in the layer. When the

Materials Research Forum LLC
https://doi.org/10.21741/9781644900475

difference in lattice constants is large, or if the symmetries of layer and substrate are different, the orientation relationship is again assumed to be predicted by the coincidence site lattice principle.

Problems appear if the structures of the layer and substrate have differing symmetries. During the initial stages of epitaxy, nuclei of the new phase are independent of each other and the displacements and orientations are governed only by the surface of the substrate. It is unclear therefore whether monocrystalline layer growth will be possible, and whether those nuclei will coalesce without creating defects.

For a given degree of in-plane lattice mismatch between a two-dimensional epitaxial layer and a substrate, there is a critical thickness above which defects form in order to relax the elastic strain. Two-dimensional lattice-matching conditions can be extended to three-dimensions and thus predict the critical size beyond which epitaxially encased nanoparticles could relax via dislocation formation. The critical particle-length at which defect-formation occurred is deduced by balancing the reduction in elastic energy which is associated with dislocation-creation against the corresponding increase in defect energy. Recent results[89], based upon a modified Eshelby inclusion technique for an embedded, arbitrarily-faceted nanoparticle, offered novel insights into the nano-epitaxy of low-dimensional structures such as quantum dots and nano-precipitates.

Analysis of the relationship between the space-symmetry groups of the new and old phases permits the determination of the conditions required for defect-free coalescence, or at least to predict the type of inter-grain boundary which will appear if defect-free coverage is not possible.

It was assumed that, if no external anisotropic effects are operating during epitaxial growth, the normal crystal structure of the growing layer is based only on structural information from the substrate surface. This permits the application of the Curie law, that symmetry of cause should produce symmetry of effect, to the epitaxial growth process. Application of the law to two-dimensional point groups of symmetry led to the proposal that 'two-dimensional point groups of symmetry of a substrate surface should be a sub-group of the two-dimensional point groups of symmetry of the contacting surface of the layer'. This principle greatly limits the possibility of single-crystal growth and the type of orientational relationship.

Considering a specific cut of the substrate, if the two-dimensional point groups of symmetry of the substrate surface could not be a sub-group of the two-dimensional point groups of symmetry of the layer, then single-crystal growth is impossible regardless of the layer orientations. In the contrary case, single-crystal growth is possible and one of those layer planes, for which the condition is met, will be parallel to the substrate surface.

If the two-dimensional point groups of symmetry of the substrate include mirror-symmetry, then the mutual angular alignment is also prescribed. It is consequently easy to determine all of the possible two-dimensional point groups of symmetry of the layer, and the corresponding Miller indices, and thus solve the problem.

A model for the interfacial energy in crystals has been developed[90] which concentrated on the geometrical origin of cusps in the energy profile. A general class of interatomic energies was first derived which explicitly incorporated the lattice geometry of the ground state. Away from the interface, the energy was supposed to be minimized by a perfect lattice. This supposition was incorporated into the energy analysis by matching locally, as far as possible, a perfect lattice to the atomic positions. The local energy was then quantified in terms of any remaining mismatch, and lattice-matching was used to derive the resultant interatomic energy. A simpler rigid-lattice model was further proposed in which the atomic positions on each side of the interface coincided with perfect but misoriented lattices. The lattice-matching was also limited to a binary choice between perfect lattices on both sides of the interface. Finally, a bound was placed on the interatomic energy and this bound was then used as a basis for comparison with experiment. Specifically considered were symmetrical tilt grain boundaries, symmetrical twist grain boundaries and asymmetrical twist grain boundaries in face-centered cubic and body-centered cubic crystals. There was found to be very good general agreement between the predicted interfacial energy structure and that calculated using molecular dynamics methods. The positions of cusps were well-predicted, as were the predominant orientations of facets.

A recently proposed crystal-matching method[91] permits the combining of any 2 crystals and the prediction of that interface structure which offers a low-strain epitaxial state. It is based only upon geometrical choices made among the possible surface cells of the 2 crystals and identifies those interfaces for which the strain, and the size of the coincidence interface cell, is small. Although those factors alone do not ensure the existence of a stable interface, simple geometrical rules make for a reasonably good point of departure.

In this approach, all of the possible crystal orientations and all of the surfaces are examined at the same time, and a scaling factor is introduced which is adjusted so as to change gradually the size of one of the matching crystal structures while the other one is kept fixed. This is rather like the real situation which exists when the lattice parameters of a material are modified by alloying. An analytical relationship between the scaling parameter and the resultant strain is then established. The scaling parameter unsurprisingly makes the method especially suitable for investigating interfaces which involve at least one alloy.

The method makes it possible to identify interfaces which have small coincidence cells and with which only a small strain is associated. It takes account only of the nature of the lattice and ignores atomic factors. The degree to which the interface is stable depends greatly upon the material properties and the nature of the bonding. To a first approximation, the situation is considered in which 2 materials are homogeneously strained and brought together so as to form an interface. The energy difference which is associated with the straining of the materials and the formation of the interface can be divided into 3 contributions. One is an interface term which represents the energy gained by forming the interface from surfaces under a given strain. The second term, which is generally small, represents the change in the surface energy which arises from the strain, and can be positive or negative. The third term is the extra energy which is associated with straining the bulk materials.

If the dependence of the surface energy upon strain can be assumed to be negligible, a stable interface is expected to exist when the energy gain arising from the interface dominates the energy expended on straining the bulk materials. The model then equates interface matches with small low-strain interface coincidence cells. Small strains can be expected to minimize the bulk strain-energy, because this scales quadratically with the strain, but it is less clear whether a small interface cell size corresponds necessarily to strong interface bonding. The latter can depend very much upon the atomic structure at the interface, and this is ignored by the model. On the other hand, when the cell-size is small, relative translation of the surfaces makes it possible to improve the bonding configuration; an improvement which then obviously extends to the entire interface and leads to high stability. If the interface cell size is large, the bonding configuration can vary greatly within the cell; with some parts being favorable to bonding, and others not, thus leading to weaker bonding.

One advantage of a small cell-size is that such cells can be more stable with respect to interface shear, given that a small cell-size tends to be associated with a marked corrugation of the energy landscape as a function of the relative interface displacement. The bonds which cross the interface will thus respond in unison to the shear and lead to a large effect. In the case of a large cell, comprising both strong and weak bonds, the latter will respond differently to shear and be associated with smaller energy corrugations.

If the materials are thick, strains have to be very small because the bulk strain-energy is proportional to the thickness, and only very low-strain interface matches are acceptable. It can also lead to the creation of incommensurate interfaces, or of defects, close to the interface and such a simple model cannot take account of those. If the interface bonding energy is very small - as in the case of van der Waals forces - and has a small energy-landscape corrugation, stable interfaces which exhibit large moiré-pattern coincidence cells or incommensurate cells can occur.

The basic method, beginning with the two-dimensional cells of the 2 crystal surfaces, is to take the three-dimensional vectors which define those surface cells and project them from three-dimensional space to two-dimensional space. Assuming the surface cell of one crystal to be defined by vectors, $\mathbf{u}_1$, $\mathbf{u}_2$, and that $\mathbf{v}_1$ and $\mathbf{v}_2$ define the surface cell of the second crystal, an affine transformation is found which maps $[\mathbf{u}_1,\mathbf{u}_2]$ into $[\mathbf{v}_1,\mathbf{v}_2]$. This transformation is then decomposed into the product of an orthonormal matrix and a positive definite symmetrical matrix. This polar decomposition yields a matrix which defines a counter-clockwise rotation of the $[\mathbf{u}_1,\mathbf{u}_2]$ onto $[\mathbf{v}_1,\mathbf{v}_2]$. With a little more manipulation, this gives the strain-matrix for any given combination of cells.

The method was applied to the semiconducting alloys, $InAs_{1-x}Sb_x$ and $Ga_xIn_{1-x}As$, and their matches to various metals. The alloys had a zincblende structure and the lattice constants were described by $a_{InAsSb} = 6.0583 + 0.4207x$ and $a_{GaInAs} = 6.0583 - 0.405x$, where x was the mole fraction of antimony or arsenic. Using the known lattice constants of the metals, the metal surfaces were strained so as to match the alloy surface. The maximum vector-length, the maximum area and the strain threshold were taken to be 50Å, 200Å^2 and 2%, respectively. A limit was also put on the Miller indices of the crystal surfaces: if the highest value of the Miller index was above a threshold value of 3, the match was discarded. The matches were calculated for a single value of the mole fraction, and relationships based upon the strain tensor were used to deduce values for other mole-fraction values.

Related calculations were made[92] of strains and stresses and of the density of elastic energy involved in these arrangements. Linear continuous anisotropic elasticity theory was used, under the assumptions: that the heterostructure was in a pseudomorphic state and free from topological defects such as dislocations at the interface, that the mismatch of the lattice parameters of the epitaxial layer and substrate was small, that bending of the structure was negligible, that the elastic properties and lattice parameters did not change across the epitaxial layer and that the interface was planar.

The layer and substrate lattices were linked mathematically by a linear transformation in which, for the 2 crystallographic coordinate systems, an arbitrary vector could be described by 2 sets of corresponding coordinates. The above assumption of a defect-free interface was then equivalent to the coincidence of the latter coordinates for any vector belonging to the interface. In the case of solid solutions, the elastic moduli and lattice parameters were taken to be linear functions of the composition.

Recalling that there are many inequivalent ways of selecting coordinate systems on the basis of translation vectors, and that choice corresponds only to the tabulated crystallographic axes, the selection is here a matter of convention. The number of settings for a given lattice is governed by its symmetry, ranging from 24 for a triclinic system to 1 for a cubic system. The misfit is also different for the various possible relationships, and should be calculated for all of them. In practice, only those epitaxial relationships having the minimum misfit, and therefore minimum elastic energy, will be observed.

A computer program called CellMatch has been developed[93] for predicting the behaviour of epitaxial layers, or heterostructures, involving van der Waals materials such as graphene. As usual, in order to analyze theoretically the smallest possible heterostructure made up of 2 van der Waals materials a common super-cell is sought. This in turn involves an energy minimization which is a compromise between the forces bonding the layers of material, and the stresses due to the differing lattice-constants. Van der Waals materials are stuck to surfaces by weak forces and thus tend to grow incommensurately with the substrate while suffering small strains within the layer.

Starting with 2 given unit-cells, a search is made for the common super-cell into which atomic structures defined by the original unit cells can be fitted so as to suffer the smallest possible strain. Taking, for example, graphene and a vicinal (332) iridium surface the fitting can be hard to achieve. The unit cells used by the program must be of slab type, having one unit-cell vector which is perpendicular to the plane spanned by the other two. Starting with the in-plane unit-cell vectors of the two unit-cells, $\mathbf{a}_1$, $\mathbf{a}_2$; $\mathbf{b}_1$, $\mathbf{b}_2$, the program first searches for common vectors, $\mathbf{v}_1$ and $\mathbf{w}_2$, in the plane to within a given tolerance. The common vectors are generated as combinations of the unit-cell vectors, $\mathbf{v}_1 = n_1\mathbf{a}_1 + n_2\mathbf{a}_2$; $\mathbf{w}_2 = m_1\mathbf{b}_1 + m_2\mathbf{b}_2$, where the indices, n_1 etc., range over all possible combinations of integers (up to 10). If the two common vectors generated by different unit cells, $\mathbf{v}_1$ and $\mathbf{w}_2$, are sufficiently close: that is, if the distance between end-points of the vectors, divided by the vector-length is less than 1%, they are short-listed as being vectors of a possible common super-cell.

Given the set of common vectors, all of the possible combinations of the two sets of common vectors $(\mathbf{v}_1, \mathbf{w}_1)$, $(\mathbf{v}_2, \mathbf{w}_2)$ which span the common super-cell are generated. All of the pairs of super-cells, with one super-cell being generated from unit-cell 1 and another from unit-cell 2, are then considered as being the common super-cell, where one super-cell is deformed so as to fit the other perfectly. Calculation of the strain is performed using standard unit-cell strain methods. All of the results are then sorted in order of increasing strain. In order to identify the energetically most favorable relative position, the space of possibilities still has to be carefully explored by considering relative shifts and rotations of the layers.

In a further mathematical refinement, the lattice-matching of two-dimensional lattices has been treated[94] in terms of applying the group isomorphism of two-dimensional Euclidean space to the complex plane. This avoided the inconveniences which arise from the algebraic structure of two-dimensional vectors. Lattice-matching was then closely related to the ideal class group, which is an invariant in the algebraic number field. An algorithm was derived for constructing a model for a superstructure formed by overlapping 2 two-dimensional lattices. This was helpful for testing trial models.

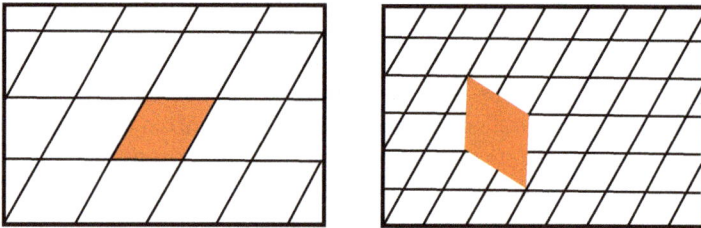

Figure 3. Lattice-match of V₃Si (111) on silicon (111)
Three silicon unit-cells (right) match, to within 0.4%, the V₃Si unit-cell (left)

Lattice-Matching to Silicon

Because silicon is such an important material, it will be useful to survey how the various lattice-matching procedures have been applied to it. The above pioneering method was, for example, later used[95] to search for all of the transition-metal silicides that geometrically lattice-match the (100), (110) or (111) faces of silicon. Transition-metal silicides constitute a special class which has a definite orientational relationship with a silicon substrate. A silicide is expected to grow epitaxially on silicon if the crystal structures are similar and their lattice mismatch is small. It seems that essentially all transition-metal silicides can be grown epitaxially on silicon to some extent.

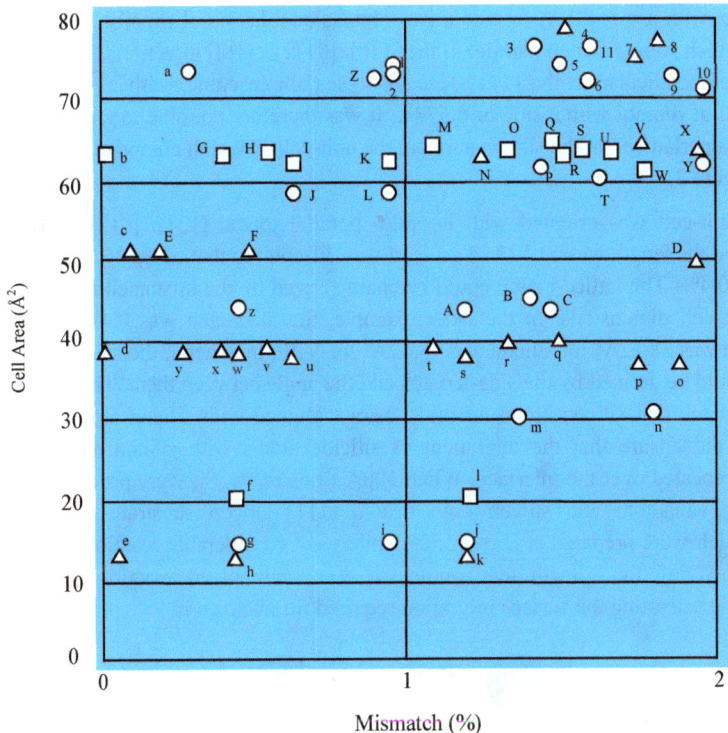

Figure 4. Lattice-matches of transition-metal silicides to silicon planes
Circles: (100), squares: (110), triangles: (111)

Key. a: $LaSi_2$, b: $RuSi$, c: $CrSi_2$, d: $RuSi$, e: Y_3Si_5, f: $NiSi_2$, g: $NiSi_2$, h: $NiSi_2$, i: $FeSi_2$, j: $CoSi_2$, k: $CoSi_2$, l: $CoSi_2$, m: Mn_4Si_7, n: Ru_2Si_3, o: Pd_2Si, p: $CrSi$, q: $ReSi$, r: Fe_5Si_3, s: $CoSi_2$, t: $TcSi$, u: $PtSi_3$, v: $OsSi$, w: $NiSi_2$, x: V_3Si, y: Ni_5Si_2, z: Nb_4Si, A: Pt_3Si, B: Rh_5Si_3, C: YSi, D: Nb_5Si_3, E: $ReSi_3$, F: $NiSi$, G: V_3Si, H: $OsSi$, I: $RhSi$, J: Nb_4Si, K: $FeSi_2$, L: $FeSi_2$, M: $TcSi$, N: $FeSi_2$, O: Fe_5Si_3, P: Y_3Si_5, Q: Rh_2Si, R: $TiSi_2$, S: $CoSi_2$, T: Y_3Si_5, U: Fe_5Si_3, V: $CrSi$, W: $TiSi_2$, X: , Y: Ni_3Si, Z: Tc_4Si, 1: Sc_2Si_3, 2: $FeSi_2$, 3: Ta_4Si, 4: Rh_5Si_3, 5: Zr_2Si, 6: Zr_2Si, 7: $NiSi$, 8: Tc_4Si, 9: Rh_3Si_2, 10: Nb_3Si, 11: $RuSi$

The lattice translations of (111) V_3Si, for example, on the (111) face of silicon (face-centerd cubic, with a = 5.431Å) formed a two-dimensional lattice of rhombi, with sides of 3.840Å, which were oriented parallel to the [1$\bar{1}$0], [01$\bar{1}$] or [$\bar{1}$0l] directions. The lattice translations perpendicular to the (111) face of V_3Si (simple cubic, with a = 4.722Å) formed a grid of rhombi with a side of 6.678Å. It was therefore possible to envisage the placing of a superlattice, on the silicon surface, the unit-cell of which comprised 3 silicon unit-cells (figure 3).

The larger unit-cell was oriented with its sides parallel to the [1$\bar{2}$1], [11$\bar{2}$] or [$\bar{2}$11] directions, and the length was 6.652Å. That is, it was different to that of the V_3Si unit cell by less than 0.4%. The lattice-match could be characterized by the mismatch and by the common unit-cell dimensions. In the latter example, the mismatch was 0.4% and the common area was 38.32Å^2 for silicon and 38.62Å^2 for V_3Si. In general, the shape (here a rhombus) would be defined by the side-lengths and the angle between them. This method did not guarantee that silicon atoms in the substrate aligned with silicon atoms in the silicide, but did ensure that the alignment of silicide atoms with silicon atoms was periodically repeated over the interface. When 500Å films of $V_{1-x}Si_x$ were produced[96] by co-depositing vanadium and silicon onto heated (111) silicon substrates, films of $V_{0.75}Si_{0.25}$ which were prepared at above 550C underwent considerable reaction with the substrate. Lowering the growth temperature to 400C minimized reaction with the substrate while activating the surface migration required for nucleation.

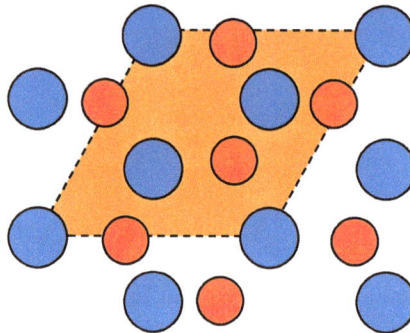

Figure 5. Lattice-match between (111)V_3Si and (111)Si structures, with layers of silicon (blue) and vanadium (red) atoms. The fundamental surface unit cell of V_3Si is essentially the same size as the ($\sqrt{3}$ x $\sqrt{3}$)R30° surface cell of (111)Si

Materials Research Forum LLC
https://doi.org/10.21741/9781644900475

In other work, every known transition-metal silicide was considered and an attempt was made to match each one to the (100), (110) or (111) faces of silicon. When the agreement was limited to a mismatch of 0.5% and a cell-area of $50Å^2$, only 5 silicides remained: V_3Si (simple cubic, a = 4.722Å), Ni_3Si_2 (trigonal, a = 5.617Å, α = 50°50'), $NiSi_2$ (face-centerd cubic, a = 5.407Å), Y_3Si_5 (hexagonal, a = 3.842Å, c = 4.144Å) and RuSi (simple cubic, a = 4.703Å).

The stresses which develop during silicide formation via metal/silicon reaction tend to be compressive, although tensile stresses can occur; perhaps due to the effect of impurities. The presence of compressive stresses shows that the overall volume decrease which accompanies silicide formation does not play any great role in stress development. When the agreement required was relaxed to a mismatch of up to 3% mismatch and a cell-area of $150Å^2$, many more matches were found. The intermediate requirements of a mismatch of 2% and a cell-area of $80Å^2$ yielded the matches shown in figure 4. The density of cases in the upper-right corner is naturally due to the doubly-poor matching: large mismatch and large unit-cell area. The apparent density is actually lower than the true one because only the matches having the smallest cell-area are included. It was noted that, of the 7 transition-metal silicides ($NiSi_2$, $CoSi_2$, Pd_2Si, PtSi, Pt_2Si, $FeSi_2$, $CrSi_2$) which were then known to grow epitaxially on silicon, 3 of them ($NiSi_2$ with a 0.4% mismatch, $CoSi_2$ with a 1.2% mismatch and Pd_2Si with a mismatch of 1.8 to 2.3%), were predicted to give a good lattice-match and grow as single crystals. The others, PtSi and Pt_2Si, grew epitaxially but were polycrystalline; the mismatch between Pt_2Si[100] and Si[110] being 2.4%, and that between PtSi[001] and Si[110] being 3.0%. It was concluded that lattice-match criteria alone were unable to predict which silicide films would grow epitaxially, even though there was a correlation between lattice-match and epitaxy. On the other hand, the case of silicon (100) on sapphire (1T•2), with its enormous mismatch of 12.5%, indicated that a good lattice-match was not necessarily essential for the occurrence of monocrystalline film.

In one particular case, superconducting thin (500-1000Å) films of V_3Si were grown[97] onto (111) silicon substrates via molecular beam epitaxy. The lattice match between (111) V_3Si and the (111) silicon surface was such that the fundamental surface unit cell of V_3Si had almost exactly the same size as that of the ($\sqrt{3}$ x $\sqrt{3}$)R30° surface cell on (111) silicon (figure 5). Because only one third of the silicon atoms lay beneath the silicon atoms of the V_3Si however, there existed 3 translationally equivalent lattice matches. For each of these there existed a pair of 180° rotation twins, thus making a total of 6 possible epitaxial matches of V_3Si to a given (111) silicon surface.

The crystallography of deposits on monocrystalline substrates was originally studied using linear transformation theories, and the two-dimensional invariant line criterion was quite successful in rationalizing the epitaxial orientations of most metal/metal systems. The lattice potential energy model was later found[98] to duplicate the predictions of invariant line criteria with regard to the case of silicides on the (001) plane of silicon. Unlike geometrical models, which ignore the chemical properties of atoms and consider only strain energy, the lattice potential energy model can simulate chemical bonding by modifying the Fourier series which is used to generate the potential surface. Epitaxial silicides constitute a special class because of their definite orientational relationship with respect to a silicon substrate. A given silicide is expected to grow epitaxially on the silicon if their crystal structures are similar and the mismatch is small. It seems that most transition-metal silicides can be grown epitaxially to a certain extent, including those having a fluorite structure, such as $CoSi_2$, $NiSi_2$ and metastable γ-$FeSi_2$. Studies of $CoSi_2$ layers[99], which were formed via the thermal reaction of vapour-deposited cobalt films on (100) silicon substrates, showed that a layer of CoSi first formed between the cobalt and silicon. The formation of $CoSi_2$ began later at the Si/CoSi interface. Because of the similarity in crystal structure, and the small mismatch between the silicon and the $CoSi_2$ lattices, the epitaxy of aligned (100)$CoSi_2$ is expected. In addition to the latter orientation however $CoSi_2$ exhibits several orientations, including a (110) preferred orientation. Many of the individual grains comprise sub-grains which are slightly rotated with respect to each other and which are connected by low-angle boundaries. This is generally attributed to the geometrical lattice match between $CoSi_2$ and silicon. Computer searches for epitaxial relationships between $CoSi_2$ and a Si(100) substrate are based upon strain-energy and coincidence site density criteria. For cobalt films which were deposited and annealed under vacua of 5 x 10^{-5} to 5 x 10^{-6}Torr at 600 to 850C, $CoSi_2$ type-C epitaxy on (111)Si was found[100]. Types A and B were also present, with type-B predominating over types A and C. The type-C orientational relationship was: (001)$CoSi_2$∥(111)Si; [110]$CoSi_2$∥[110]Si. Corresponding to the 3 equivalent <110> silicon directions, 3 type-C variants with structurally equivalent domains could be grown epitaxially on (111)Si (figure 6).

Only one set of parallel dislocations occurred at the C-type $CoSi_2$∥(111)Si interface, and these were of mixed type, with a Burgers vector of ½<110>. It was suggested that the type C epitaxy on (111)Si was the result of carbon and oxygen contamination of the cobalt layer. Films of $CoSi_2$, which were formed via the thermal reaction of a 120Å cobalt film on Si(001), consisted largely of epitaxial grains with various orientations[101]. Twin-oriented $CoSi_2$(111) grains grew epitaxially on Si{111} facets which appeared during annealing. Two distinct mosaic structures were observed. Epitaxial grains having the same orientation as the silicon substrate, such as $CoSi_2$(001) on Si(001) and $CoSi_2$(111) on Si{111} facets, had a mosaicity of ~0.5° full-width at half-maximum, while those of differing orientation had a mosaicity of ~2.5° full-width at half-maximum. The smaller mosaicity of epitaxial grains of the same orientation was attributed to a reduced interfacial energy due to a higher coincidence site density.

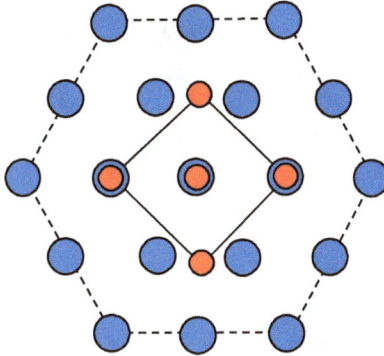

Figure 6. Projection of the CoSi₂ unit cell along the [001]CoSi₂ direction on the (111)Si plane showing one variant of CoSi₂, grown epitaxially on (111)Si, along one of the three <110>Si directions, silicon: blue, CoSi₂: red

Alloy films, $Co_{1-x}Ni_xSi_2$, were grown[102] on Si(100) via the solid-state reaction of homogeneous co-deposited $Co_{1-x}Ni_x$. It was anticipated that the change in the lattice parameter, associated with an increased nickel content, would tailor an improvement in the epitaxial quality of the film. This prediction was correct up to a concentration of 15%Ni, but even this improvement was due to a large volume fraction of (110)-oriented grains rather than a (100) orientation. It transpired that the most widespread texture was not guaranteed to be the one giving the best match to the substrate nor the one providing continuity of the atomic planes across the interface. The popular geometrical arguments could not therefore explain the extent of (110)-oriented grains. Growth kinetics also clearly played a key role during texture development. At nickel concentrations of more than 15%, the epitaxial quality sharply decreased. A polycrystalline film formed at 40%Ni. The decrease was attributed to a change, in the disilicide nucleation site, from the film/substrate interface to the film surface. At nickel concentrations above 50%, the (100) orientation predominated due to the growth of a NiSi₂-rich film at the interface.

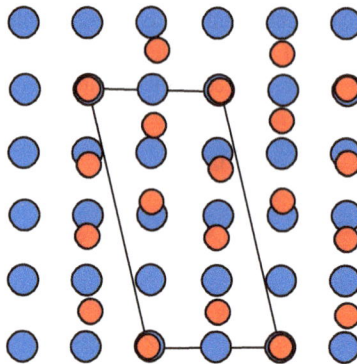

Figure 7. Atom arrangement at the CrSi₂/Si(001) interface and common unit cell
Only chromium atoms in the CrSi₂(001) plane are shown
Blue: silicon atoms, red: chromium atoms

Epitaxial $CrSi_2$, grown onto (001) and (011) silicon surfaces, was investigated[103] using transmission electron microscopy. The quality of the epitaxial $CrSi_2$ layers, with regard to the size, extent of surface coverage and regularity of interfacial dislocations in the epitaxial regions, was best for the (111) silicon surface and worst for (011). Crystallographic analysis was used to examine possible matches between atoms in the corresponding $CrSi_2$ planes and the (111), (001) and (011) Si planes at the interfaces. The quality of the $CrSi_2$ epitaxy was found to be directly related to the lattice-match. When $CrSi_2$ was grown onto Si(001) by solid-phase reaction, transmission electron microscopy revealed[104] another epitaxial growth relationship: $CrSi_2(001)\|Si(001)$; $CrSi_2(110)\|Si(110)$. Layers of $CrSi_2$ were grown[105] onto Si(001) by molecular beam epitaxy using the template technique, with the nominal thickness of chromium required to create the template ranging from 0.2 to 0.5nm. The greatest degree of texture was found for silicide films that had been formed by depositing 0.4nm of chromium. This film consisted of regions with two morphologies which corresponded to 2 different epitaxial orientations (figures 7 and 8). Most of the crystallites grew with: $CrSi_2(001)[100]\|Si(001)[110]$. Smaller portions of the layer consisted of crystallites having the orientation: $CrSi_2(112)[1\bar{1}0]\|(001)[110]$. Two perpendicular domains were observed for each orientation. This could be explained in terms of the crystal symmetry of the film and the substrate.

Perfect Czochralski silicon tetra-crystals and bi-crystals were investigated[106] with regard to the interaction of thermally-generated lattice dislocations with $\Sigma = 9$ boundaries, and the structural anomalies of the latter boundaries in tetra-crystals with a [221] growth-axis. The reflection, absorption and emission of dislocations at the boundary could all be explained in terms of the coincidence site and displacement shift complete geometrical lattice models for a diamond-type structure.

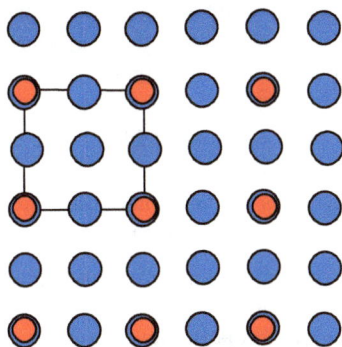

Figure 8. Alternative atom arrangement at the CrSi₂/Si(001) interface
Only chromium atoms in the CrSi₂(001) plane are shown
Blue: silicon atoms, red: chromium atoms

Solid-phase, and reactive deposition, epitaxy have recently been used[107] to grow islands which were crystallographically identical, but morphologically different. Tetragonal α-FeSi₂ islands crystallized pseudomorphically with α-FeSi₂(112)<110>||Si(111)<110> and α-FeSi₂ (110)<111>||Si(220)<112> orientations and flat (2 x 2)-reconstructed top facets. The solid-phase epitaxially grown islands self-ordered in the form of one-dimensional chains which decorated the vicinal Si(111) step bunch edges along a specific $1\bar{1}0$ direction. The reaction-deposited islands elongated along 3 equivalent $11\bar{2}$ directions, and tended to be shorter, narrower and thinner than the other islands. When epitaxially aligned films of β-FeSi₂ were grown[108] onto (001) silicon by reactive deposition epitaxy, molecular-beam epitaxy and solid-phase epitaxy, the matching crystallographic faces, FeSi₂(100)||Si(001), remained the same while 2 different azimuthal orientations predominated; depending upon the deposition method and the growth temperature. Films having the FeSi₂[010]||<110> orientation, and grown using reactive deposition at 500C, were monocrystalline and of large area. This hetero-epitaxial relationship had a common unit mesh area of 59Å^2, with a mismatch of 2.1%[109]. There is a strong tendency toward island formation within this hetero-epitaxial system. The surface morphology was rough however due to islanding which preceded the formation of a continuous film. Films of azimuthally-oriented FeSi$_2$[010]∥Si<100>, grown by solid-phase epitaxy at 250C or by molecular beam epitaxy at down to 200C, exhibited a much smoother surface morphology.

Early work[110] on the TiSi$_2$/Si system, involving high-resolution transmission electron microscopy and electron diffraction, showed that large crystallographic differences between the crystals, and a complex reaction-path, did not prevent the formation of flat well-defined interfaces. The ($\bar{1}$01) plane of TiSi$_2$ was found to be a preferred plane for epitaxial growth on the (111) plane of silicon (table 2). The terminal TiSi$_2$ plane at the interface there comprised a single atomic species. This matching was attributed to the small discrepancy in atomic densities and to the interplanar spacings. The local epitaxial relationships minimized the two-dimensional misfit at the interface (figure 9).

First-principles theories of hetero-epitaxy for films with lattice misfits greater than 10% are rare because the film is strain-free away from the interface, but strains and dislocations develop at the interface. On the other hand, the films may be free from threading-dislocations because all of the dislocations are confined to the interface. Heterogeneous films have been modelled[111] by combining first-principles and elasticity theory in order to predict the epitaxial relationship. The predictions for (111)Al∥(111)Si, (111)Cu∥(111)Si, (001)Cu∥(001)Si and (111)CaF$_2$∥(001)Ni were in good agreement with experiment.

The growth of titanium and the formation of epitaxial titanium silicide on the Si(111)-(7x7) surface were later investigated[112] using reflection high-energy electron diffraction and high-resolution transmission electron microscopy. The growth-mode of the titanium was of Stransky-Krastanov type when the substrate was at room temperature, but was of Volmer-Weber type when the substrate temperature was about 550C. The results showed that C54 TiSi$_2$ grew epitaxially on a silicon substrate when 160 monolayers of the titanium were deposited onto a Si(111)-(7x7) surface at room temperature and then annealed (750C, 600s) under ultra-high vacuum. This interface tended to be rather incoherent, but the resultant TiSi$_2$ crystallites were monocrystalline, with matchings of the form: (111)TiSi$_2$∥(111)Si; (311)TiSi$_2$∥(111)Si and (022)TiSi$_2$∥(111)Si. A thin monocrystalline silicon overlayer with a [111] direction grew on the TiSi$_2$ surface when TiSi$_2$∥(111)Si was annealed at about 900C under ultra-high vacuum.

An additional epitaxial relationship, [100](02$\bar{2}$)TiSi$_2$)∥[1$\bar{1}$0](1$\bar{1}\bar{1}$)Si, was subsequently established[113] for TiSi$_2$ which was prepared by using the self-aligned silicide process on

Si(001). The silicide film was heavily strained near to grain boundaries. The above relationship also explained a preferred orientation relationship, $(040)TiSi_2\|(001)Si$, which had been reported. One prominent sign of the epitaxial relationship was a clear moiré pattern which ran through the entire interface.

Table 2. Local epitaxial relationships of $TiSi_2$ on $(1\bar{1}\bar{1})$ silicon and the relative rotation about the interface normal for $(\bar{1}01)TiSi_2$ epitaxies

Relationship	Angle (°)
$(1\bar{1}\bar{1})Si\|(\bar{1}01)TiSi_2; [110]Si\|[010]TiSi_2$	0
$(1\bar{1}\bar{1})Si\|(\bar{1}01)TiSi_2; [110]Si\|[1\bar{2}1]TiSi_2$	8.9
$(1\bar{1}\bar{1})Si\|(\bar{1}01)TiSi_2; [110]Si\|[1\bar{3}1]TiSi_2$	19.1
$(1\bar{1}\bar{1})Si\|(\bar{1}01)TiSi_2; [110]Si\|[161]TiSi_2$	22.4
$(1\bar{1}\bar{1})Si\|(\bar{1}01)TiSi_2; [110]Si\|[1\bar{4}1]TiSi_2$	28.2
$(1\bar{1}\bar{1})Si\|(\bar{1}01)TiSi_2; [110]Si\|[101]TiSi_2$	30.0
$(1\bar{1}\bar{1})Si\|(\bar{1}01)TiSi_2; [110]Si\|[141]TiSi_2$	31.8
$(1\bar{1}\bar{1})Si\|(\bar{1}01)TiSi_2; [110]Si\|[121]TiSi_2$	51.1
$(1\bar{1}\bar{1})Si\|(010)TiSi_2; [110]Si\|[001]TiSi_2$	-
$(1\bar{1}\bar{1})Si\|(2\bar{1}1)TiSi_2; [110]Si\|[1\bar{1}1]TiSi_2$	-
$(1\bar{1}\bar{1})Si\|(001)TiSi_2; [110]Si\|[100]TiSi_2$	-
$(1\bar{1}\bar{1})Si\|(100)TiSi_2; [110]Si\|[001]TiSi_2$	-

The growth of silicide islands following the reactive deposition of titanium on (111) silicon at about 850C was studied[114] using atomic force microscopy and transmission electron microscopy. Their predominant shape was very long and narrow, and they were treated as being nanowires. Most of the nanowires were oriented along <220> silicon directions, with average dimensions of 20nm in width, 10nm in height and several microns in length. The predominant nanowire structure was incommensurate and was thought to consist of C49 $TiSi_2$. A few of the nanowires were oriented along <224> silicon directions. These tended to break up into chains of small segments of regular size and spacing. Growth at lower temperatures or higher deposition rates produced smaller

Materials Research Forum LLC
https://doi.org/10.21741/9781644900475

and more numerous nanowires, their length apparently limited by intersection with other nanowires which were oriented 120° apart. Flat-topped structures co-existed with the nanowires, and included small equilateral triangles and large rectangular plates. The triangular islands consisted of fully-relaxed C54 $TiSi_2$, while the chains consisted of relaxed C49 $TiSi_2$.

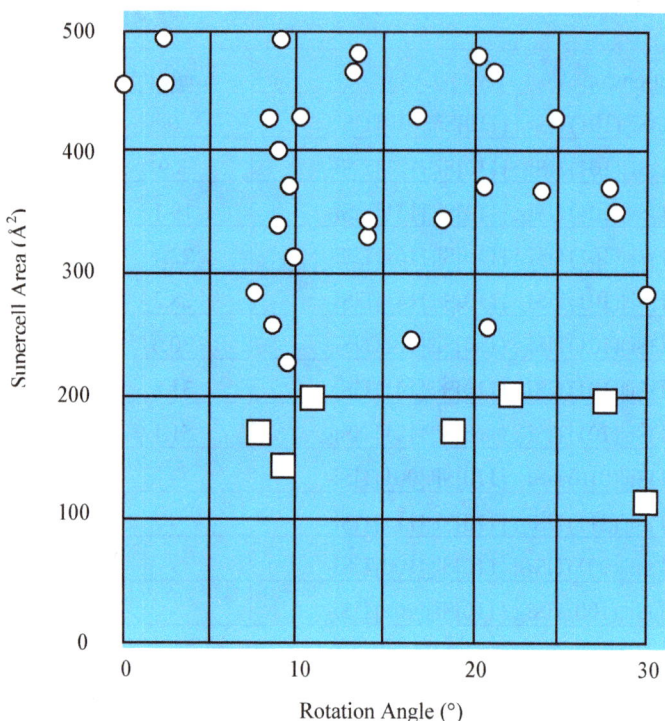

Figure 9. Areas which correspond to almost-coincident super-cells as a function of the rotation-angle between two-dimensional silicide and silicon lattices in the common interface planes, ($\bar{1}01$)$TiSi_2$ and ($1\bar{1}\bar{1}$)Si. Squares indicate those areas which are smaller than 200Å

In more recent work, titanium silicide was grown[115] on (111) silicon substrates using reactive deposition or solid-phase epitaxy. The nanocrystals which resulted from solid-state reaction were very non-uniform in shape and size, and post-deposition annealing increased that non-uniformity. The relaxation of epitaxial mismatch strain by misfit dislocations was deduced from a gradual reduction in the nanocrystal vertical aspect ratio and the development of flat-topped facets. Silicide nanocrystals which were instead formed by reactive deposition exhibited a high uniformity and thermal stability. Marked strain relaxation was indicated by nanocrystal flattening at temperatures above 650C, and by subsequent coalescence. Although stable C54 $TiSi_2$ eventually formed in both cases, it was concluded that superior titanium silicide nanocrystal arrays were more easily obtained by reactive deposition epitaxy.

It is interesting to compare these results with those[116] for epitaxial layers of the tetragonal and hexagonal silicide phases of the non-transition metal, molybdenum, on various silicon planes. In the case of (001) silicon, randomly oriented grains of molybdenum which were some 100Å in size were observed in as-deposited samples. The hexagonal silicide had a grain size of 150 or 350Å following annealing at 500 or 600C, respectively. In samples which were heat-treated at 700 to 1000C, the molybdenum grains were entirely transformed into $MoSi_2$, with the tetragonal form predominating. The average grain-sizes were 0.1. 0.15, 0.2 and 0.3μm during subsequent annealing at 700, 800, 900 and 1000C, respectively.

Three forms of the tetragonal silicides were found upon viewing along [001]Si: (A) [110]$MoSi_2$∥[001]Si; $(00\bar{4})MoSi_2$∥$(2\bar{2}0)$Si, (B) [111]$MoSi_2$∥[001]Si; $(11\bar{2})MoSi_2$∥$(2\bar{2}0)$Si, (C) [100]$MoSi_2$∥[001]Si; (004)$MoSi_2$∥(220)Si. Forms A and B were often associated with each other in the same region. The interfacial dislocations at $MoSi_2$∥Si interfaces had the same configuration for all 3 forms: edge-type with a Burgers vector of ½<110> and a spacing of about 90Å. For the A and B forms, the atomic arrangements on (110) and (116) planes, normal to the [110] and [111] directions, respectively, of the tetragonal silicide, were similar to that on the (001) silicon plane. The lattice mismatches along the [220] and $[2\bar{2}0]$ directions of (001)Si for type-A and type-B silicides were 2.28 and -1.69% - and 2.19 and -1.69% - respectively.

The Burgers vectors of interfacial dislocations were parallel to those directions exhibiting large positive mismatches of 2.28 and 2.19% for forms A and B, respectively. Annealing at high temperatures reduced and increased, respectively, the lattice mismatches along directions exhibiting negative and positive values. It was deduced that interfacial dislocations were more likely to be generated with Burgers vectors lying along directions having large positive mismatches than along those having smaller negative mismatches. In type-C epitaxial silicide, the lattice mismatches along the [220] and $[2\bar{2}0]$ directions were 2.28 and 0.09%, respectively, at room temperature. The Burgers vectors of the interfacial dislocations were parallel to the direction of large positive mismatch: [220]. Type-C epitaxial silicide was much less often observed than were type-A and type-B epitaxial silicides.

Two forms of the hexagonal silicide grew on (001)Si: (A) [00•1]MoSi$_2$∥[001]Si; (20•0) MoSi$_2$∥(2$\bar{2}$0)Si, (B) [$\bar{2}$4•3]MoSi$_2$∥[001]Si; (2$\bar{1}$•2)MoSi$_2$∥(2$\bar{2}$0)Si. Form-A of the hexagonal silicide was often associated with form-A of the tetragonal silicide, within the same region. The mismatches at [00•1]MoSi$_2$∥[00l]Si were 4.02 and 0.11% along the [220] and [2$\bar{2}$0] directions of (001)Si, respectively, at room temperature. The straight interfacial dislocations were of edge-type, with a ½[110] Burgers vector, and the average spacing was about 90Å. The Burgers vector was again parallel to the direction of large positive mismatch. The difference between the hexagonal and tetragonal silicides was due only to their stacking order: tetragonal ABAB and hexagonal ABCABC. Because of the very small difference in free energy between the tetragonal and hexagonal phases, faults in stacking could occur during growth, giving overlapping epitaxial hexagonal and tetrahedral material or vice versa.

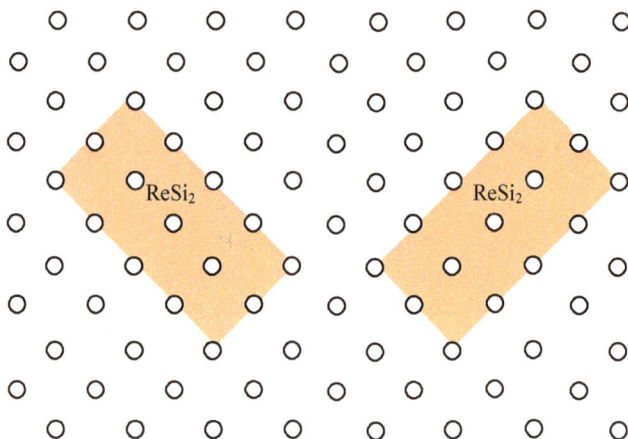

*Figure 10. Predicted common unit cell for ReSi$_2$ (100) on the Si(001) face
Note the two distinct but equivalent azimuthal orientations of the cell*

Materials Research Forum LLC
https://doi.org/10.21741/9781644900475

The mismatches between ($\bar{1}2 \bullet 2$) hexagonal $MoSi_2$ and (001)Si, along the [400] and [040] directions of silicon were -1.92 and -2.90%, respectively. The mismatches were considerably smaller in magnitude at high temperatures than at room temperature. The minimum spacing was over 50Å. The variants of type-A and type-B epitaxy were oriented relative to each other by a 90° rotation around the [001]Si axis. The silicide grains had a tendency to rotate towards directions which had the greatest epitaxial relationship with respect to the silicon.

Epitaxial hexagonal silicides with [00$\bullet$1]$MoSi_2$||[111]Si and tetragonal silicides with [111]$MoSi_2$||[111]Si or [110]$MoSi_2$||[111]Si grew in (001) wafers. In the case of [111]$MoSi_2$||[111]Si, tetragonal [11$\bar{2}$]$MoSi_2$ was parallel to (20$\bar{2}$)Si. About 50% of the silicide had an epitaxial relationship with the silicon substrate, but the fraction of hexagonal epitaxial $MoSi_2$ was by far the smaller. All 5 forms of epitaxy occurred on (111)Si wafers. Epitaxial $MoSi_2$ also grew on (011)Si wafers. The fact that epitaxial hexagonal $MoSi_2$ was often present in samples annealed at 1100C suggested that it was more stable.

Transmission electron microscopy of iridium thin films on silicon revealed[117] the formation of precursor phases, IrSi and $IrSi_{1.75}$, at 300 to 500 and 600 to 900C, respectively. Stable $IrSi_3$ grew epitaxially on (111) and (001) silicon faces at 1000 to 1100C, with 3 predominant modes on (111) and a single dominant mode on (001). The 3 orientation relationships between epitaxial silicide and (111)Si were: (A) [2$\bar{1}\bullet$0]$IrSi_3$||[111]Si; (01$\bullet$0)$IrSi_3$||(2$\bar{2}$0)Si, (B) [00$\bullet$1]$IrSi_3$||[111]Si; (10$\bullet$0)$IrSi_3$||(2$\bar{2}$0)Si, (C) [03$\bullet\bar{1}$]$IrSi_3$||[111]Si; ($\bar{1}$0$\bullet\bar{3}$)$IrSi_3$||(2$\bar{2}$0)Si. In case A, regularly spaced (24nm) interfacial dislocations of edge or mixed type were found, with a ½<1$\bar{1}$0> Burgers vector. In case B, there was incommensurate matching and no interfacial dislocations were detected. In case C, regular edge or mixed interfacial dislocations with a spacing of 48nm were identified which had a 1/6<11$\bar{2}$> Burgers vector. The areal fractions which were occupied by the 3 forms of epitaxy were almost the same, although some 10% were of the form: (D) [10$\bullet\bar{2}$]$IrSi_3$||[111]Si; (01$\bullet$0)$IrSi_3$||(2$\bar{2}$0)Si. No interfacial dislocations were found in those regions. The best results were found for (111) samples annealed at 1000C. Regions of epitaxial $IrSi_3$ of up to 40μm in size occurred on (111). The Burgers vectors of edge-type dislocations tended to lie along directions of greater lattice mismatch. A superlattice structure of $IrSi_3$ existed which was hexagonal and had a unit-cell size that was 3 times larger than that of $IrSi_3$. There was no direct correlation between the degree of lattice-match and the epitaxial quality.

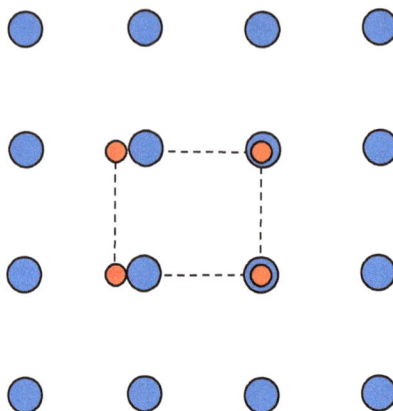

Figure 11. Epitaxial growth of (1T•0)YSi$_{2-x}$ film on the (100) plane of silicon (1T•0)YSi$_{2-x}$||(100)Si; [TT•0]YSi$_{2-x}$||[01T]Si. Mismatch: 0.05%
Silicon: blue, yttrium or silicon in YSi$_2$: red

Thin ReSi$_2$ films were grown[118] on (001) silicon wafers via the vacuum-evaporation of rhenium onto hot substrates under ultra-high vacuum. The preferred epitaxial relationship (figure 10) was (100)ReSi$_2$||(001)Si; [010]ReSi$_2$||<110>Si. The lattice-matching involved a common unit cell which was 120Å^2 in extent, with a mismatch of 1.8%. There was a very high degree of alignment between the ReSi$_2$ (100) and silicon (001) planes. Transmission electron microscopy further revealed the presence of rotation twins, having lateral dimensions of the order of 100Å, which corresponded to 2 distinct but equivalent orientations of the common unit cell.

The domain structure in YSi$_{2-x}$ films grown on (100) silicon substrates by solid-state reaction at above 400C was determined[119] by using X-ray diffraction methods. Films which were produced using on-axis substrates and 2°-tilted substrates had a preferred orientation, with the (1T•0) plane of the silicide lying parallel to the (100)Si surface. The films comprised 2 types of domain: [00•1]YSi$_{2-x}$ being parallel to [01T]Si or [0TT]Si. Rocking curve measurements of the (2Z•T)YSi$_{2-x}$ asymmetrical reflection showed that the volume fractions of the 2 types of domain were almost equal on the on-axis (100)Si substrate and differed by about 30% on the tilted substrate. The origin of the double-domain structure was analyzed on the basis of geometrical matching at the interface between (1T•0)YSi$_{2-x}$ film and the (100)Si substrate (figures 11 and 12).

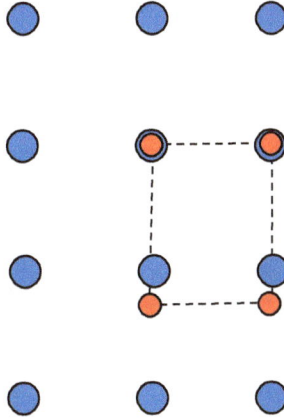

Figure 12. Epitaxial growth of $(1\overline{1}\bullet0)YSi_{2-x}$ film on the (100) plane of silicon $(1\overline{1}\bullet0)YSi_{2-x}||(100)Si;\ [00\bullet1]YSi_{2-x}||[0\overline{1}\overline{1}]Si.$ Mismatch: 7.9% Silicon: blue, yttrium or silicon in YSi_2: red

The epitaxial growth of tetragonal and hexagonal WSi_2 on the (001), (111) and (011) planes of silicon, studied[120] using transmission electron microscopy, involved 8 different forms of epitaxy.

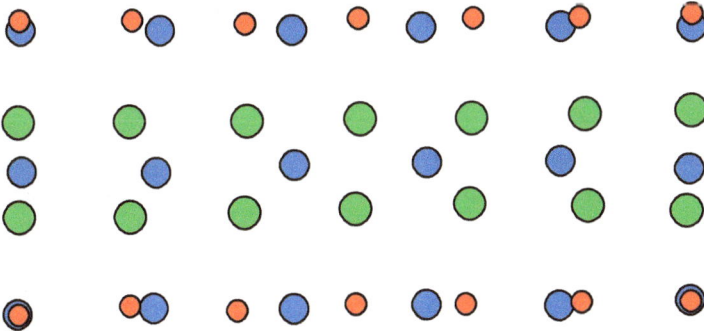

Figure 13. Matching of atoms of (100) plane of tetragonal WSi_2 with (001) silicon substrate

Silicon atom in silicon: blue, silicon atom in WSi$_2$: green, tungsten atom in WSi$_2$: red.

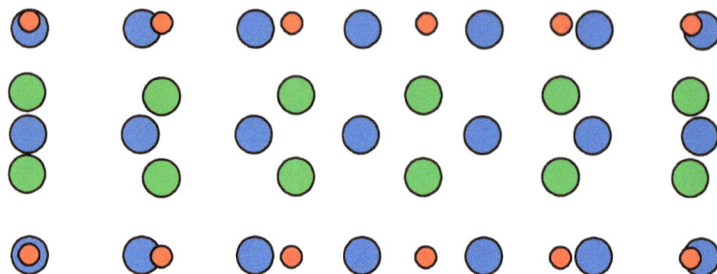

Figure 14. Matching of atoms of (116) plane of tetragonal WSi$_2$ with (001) silicon substrate. Silicon atom in silicon: blue, silicon atom in WSi$_2$: green, tungsten atom in WSi$_2$: red.

Crystallographic analysis determined the possible matches which could occur, between atoms in the overlayer and silicon, at the interfaces. Three modes of epitaxial growth of tetrahedral WSi$_2$ were identified. These were: (A) [110]WSi$_2$||[001]Si; (00$\bar{4}$)WSi$_2$||(2$\bar{2}$0)Si, (B) [111]WSi$_2$||[001]Si; (11$\bar{2}$)WSi$_2$||(2$\bar{2}$0)Si, (C) [100]WSi$_2$||[001]Si; (004)WSi$_2$||(220)Si. Modes A and B were more frequently observed than was mode-C, and tended to be connected with each other. The atomic arrangements of tetrahedral WSi$_2$ on the (110) and (116) planes, normal to [110] and [111], respectively, were close to those on the (001) silicon plane (figures 13 and 14). For all 3 models of epitaxial tetrahedral WSi$_2$ on (001) silicon, the interfacial dislocations were of edge type, with ½<110> Burgers vectors and an average spacing of about 80A. In the case of (111) samples, straight dislocations with Burgers vectors parallel to directions having larger positive mismatches than other directions were observed. For mode-A and mode-B, the mismatches were 2.47% and -1.43%, and 2.47% and -1.43%, along the [110] and [1$\bar{1}$0] silicon directions, respectively, at room temperature and the Burgers vector of the dislocations lay along the [110] direction. For mode-C epitaxy, the mismatches were 2.47% and 0.37% along the [110] and [1$\bar{1}$0] silicon directions of the substrate. The Burgers vectors again lay parallel to directions of large positive mismatch. Two forms of hexagonal WSi$_2$ epitaxy on (001)Si were found (table 3): (A) [00•1]WSi$_2$||(001]Si; (20•0)WSi$_2$||(2$\bar{2}$0)Si and (B) [$\bar{2}$4•3]WSi$_2$||[001]Si; (2$\bar{1}$•2)WSi$_2$||(2$\bar{2}$0)Si. Mode-A here often generated neighboring mode-A epitaxial tetrahedral WSi$_2$ on (001)Si.

The interfacial dislocations were edge-type with a ½<110> Burgers vector and an average spacing of about 80Å. The lattice mismatches were 4.04 and 0.13% along the [1$\bar{1}$0] and [110]Si directions, respectively, and the Burgers vector of the interfacial dislocation was again parallel to the direction of large positive mismatch. For mode-B epitaxy, the lattice mismatches were -3.01% and -1.84% along the [100] and [010]Si directions, respectively. No interfacial. dislocations were observed, thus indicating that the silicide overlayer was pseudomorphic with respect to the substrate. Variants of epitaxial material, which were structurally equivalent on the basis of symmetry considerations, were frequently observed on different areas of the same sample. The variants were oriented with a 90° rotation around the [001]Si direction with respect to each other. The silicide exhibited a general tendency to rotate into orientations which were presumably also low-energy states.

Table 3. Lattice-matches of epitaxial WSi$_2$ to silicon

Mode	Matching Planes	Epitaxy	Mismatch: a(%)	b(%)	α(%)
A^t	(110)WSi$_2$‖(001)Si	[001]‖[110]	2.47	-1.43	0
B^t	(116)WSi$_2$‖(001)Si	[1$\bar{1}$0]‖[1$\bar{1}$0]	2.47	-1.43	0
C^t	(100)WSi$_2$‖(001)Si	[001]‖[110]	2.47	0.37	0
D^t	(110)WSi$_2$‖(111)Si	[001]‖[1$\bar{1}$0]	2.47	2.47	0
E^t	(116)WSi$_2$‖(111)Si	[1$\bar{1}$0]‖[$\bar{1}\bar{1}$2]	2.47	-2.46	0
A^h	(00•1)WSi$_2$‖(001)Si	(2$\bar{1}$•0)WSi$_2$‖(1$\bar{1}$0)Si	4.04	0.13	0
B^h	($\bar{1}$2•2)WSi$_2$‖(001)Si	(1$\bar{2}$•3)WSi$_2$‖(010)Si	-3.01	-1.84	0
C^h	(00•1)WSi$_2$‖(111)Si	[1$\bar{1}$•0]WSi$_2$‖(1$\bar{1}$0)Si	4.04	4.04	0

a, b and α are the sides of the common unit cell and the included angle, t = tetragonal, h = hexagonal

Table 4. Lattice-matches of epitaxial Pd_2Si to silicon

Mode	Matching Planes	Area (Å^2)	Mismatch: a(%)	b(%)	α(%)
A	$(00\bullet1)Pd_2Si\|\|(111)Si$	37	-2.36	-2.36	0
B	$(1\bar{1}\bullet0)Pd_2Si\|\|(110)Si$	401	1.45	-5.25	0
C	$(\bar{2}4\bullet3)Pd_2Si\|\|(\bar{1}14)Si$	481	0.6	-1.18	-1.2
D	$(1\bar{1}\bullet0)Pd_2Si\|\|(\bar{1}11)Si$	223	-2.36	-0.84	0

a, b and α are the sides of the common unit cell and the included angle

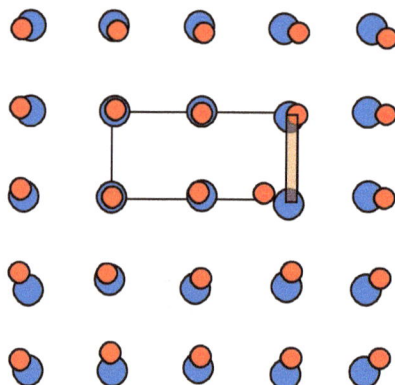

Figure 15. Lattice matching of PtSi and silicon, A orientation, showing mismatch
Red: platinum atoms, blue: silicon atoms

The epitaxial growth and thermal stability of Pd_2Si on (001), (011) and (111) silicon substrates was investigated[121] using transmission electron microscopy, X-ray diffraction and sheet-resistance methods, showing that Pd_2Si was most stable on (111). Full surface coverage occurred on annealed (800C, 1h) samples. The general trends in the thermal stability were similar for (001) and (011) samples. The agglomeration of Pd_2Si began in samples which were annealed at temperatures as low as 600C. The growth of laterally uniform Pd_2Si on (111) samples was attributed to the extensive growth of the epitaxial silicide regions. Four modes of epitaxial growth were observed. The predominant mode corresponded to good lattice-matching with the substrate. In order to calculate lattice mismatches at the $Pd_2Si\|Si$ interfaces, the silicide's hexagonal unit-cell, with a = 0.653nm and c = 0.344nm, was chosen so as to be consistent with high-resolution electron microscopic studies of the structure of the $(00\bullet1)Pd_2Si\|(111)Si$ interface.

The lattice matches for the various types of epitaxy (table 4) indicated that the common unit-cell for mode-A was relatively small. This in turn corresponded to a better lattice-match, with respect to the silicon, than did modes B, C and D. Misorientations by 4 to 7° away from the mode-A epitaxial relationship were frequent.

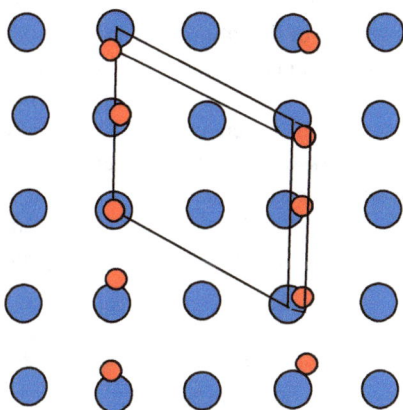

Figure 16. Lattice matching of PtSi and silicon, B orientation, showing mismatch
Red: platinum atoms, blue: silicon atoms

The epitaxial growth of PtSi on (111) silicon, investigated[122] using lattice imaging, was such that small (20 to 40nm) PtSi domains having 3 equivalent positions on Si(111) grew epitaxially all over the substrate, with a sudden transition occurring between the PtSi lattice to the silicon lattice. The interface contained atomic steps, as verified by image simulations of the interface. In various proposed models, inclined interfaces were suggested to increase the local coherency between the PtSi and silicon. The PtSi epitaxial orientation and morphology depended greatly upon the (001) silicon substrate temperature during platinum deposition[123]. When 15nm-thick platinum films were evaporated onto substrates at temperatures ranging from ambient to 1100K, the PtSi formed a continuous film below 850K and islands at above 920K. Over most of that range, the predominant epitaxial orientation was $(1\bar{1}0)$PtSi$\|$(001)Si (A, figure 15). At about 1000K, the predominant orientation was $(1\bar{2}1)$PtSi$\|$(001)Si (B, figure 16). Between 920 and 950K, the B-types grains were partially A-type. Below 1000K, the main factor which controlled the epitaxial orientation was area-mismatch: the difference between the PtSi and silicon superlattice areas at the interface. This explained the predominance of A-type grains at above 1050K.

The high-temperature epitaxy of PtSi on (001) silicon surfaces was again investigated[124], for deposition temperatures of up to 850C, by means of ultra-high vacuum transmission electron microscopy. Polycrystalline films with 3 preferred orientations were observed: [(110), (120) or (210)]PtSi$\|$(001)Si with [100] or [120]PtSi$\|$[220]Si and [220], [120] or [210] PtSi$\|$[$\bar{2}$20]Si. An alignment of the PtSi(002) planes with Si(220) planes was found at all temperatures. At 600C, continuous polycrystalline films occurred in the form of rectangular (10 x 30nm) grains with a predominant PtSi(110) preferred orientation. At 750C, the grains formed islands which were elongated in the PtSi[001] direction parallel to Si[110]. There appeared a marked anisotropy of shape, with length/width ratios of up to 100. Interface faceting was observed in all of the islands at 810C, and the preferred orientation became largely (120). An alignment of the PtSi(001) planes with Si(220) planes, giving a mismatch of 6.1% was preserved at all temperatures.

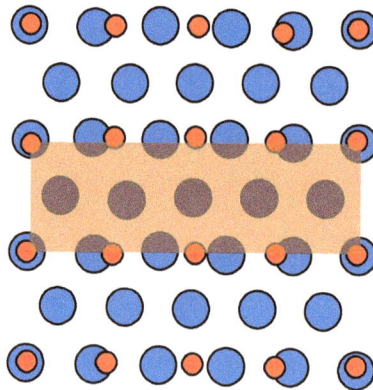

Figure 17. Interfacial matching A between NbSi$_2$ and silicon
Red: niobium or silicon in NbSi$_2$, blue: silicon in substrate

Thermodynamic data for the Ru-Si system, and computer analysis applied to the experimental phase diagram, show[125] that Ru_2Si, Ru_4Si_3, RuSi and Ru_2Si_3 can be treated as stoichiometric phases. The epitaxial phases Ru_2Si_3, RuSi and Ru_2Si were found[126,127], where Ru_2Si_3 was stable at high temperatures and resulted from long-term annealing of the other phases. Moiré fringes were observed in the case of RuSi, and interfacial dislocations were observed in the case of Ru_2Si_3 and Ru_2Si. The average interfacial dislocation spacings ranged from 1000 to 4000Å for Ru_2Si_3/Si. These large spacings indicated that the mismatch between silicide and silicon was relatively small and should promote the epitaxial growth of Ru_2Si_3 on the Si(111) surface. The orientation relationship between Ru_2Si_3 and silicon was: $[010]Ru_2Si_3\|[111]Si$; $(402)Ru_2Si_3\|(022)Si$. The cubic RuSi grew epitaxially on silicon at between 900 and 1100C. Moiré fringes with a spacing of about 80Å were observed at the RuSi/Si interface, and no interfacial dislocations were found. The orientation relationship was: $[111]RuSi\|[111]Si$; $(112)RuSi\|(202)Si$. Orthorhombic Ru_2Si grew epitaxially on (111)Si during surface during annealing at 900 to 1100C. Disconnected $1\mu m$ epitaxial regions occupied less than 5% of the surface, and the interfacial dislocations were of edge type, with a $1/6<112>$ Burgers vector and an average spacing of about 100Å. The lattice mismatch of this phase was much greater than that of the other phases. The orientation relationship was: $[110]Ru_2Si\|[110]Si$; $(004)Ru_2Si\|(220)Si$.

Transmission electron microscopy and Auger electron spectrometry revealed[128] 2 distinct forms of epitaxial growth of $NbSi_2$ on Si(111) following annealing (1100C, 1h). The orientational relationships were (A) $[1\bar{1}\bullet0]NbSi_2\|[111]Si$; $(00\bullet3)NbSi_2\|(22\bar{4})Si$ (figure 17) and (B) $[1\bar{1}\bullet0]NbSi_2\|[111]Si$; $(00\bullet3)NbSi_2\|(2\bar{2}0)Si$ (figure 18). In the case of type-A epitaxy, interfacial dislocations with a spacing of 40 to 80nm were observed which were of edge type with a Burgers vector of $1/6<112>$. No interfacial dislocations were found for type-B epitaxy. The epitaxy correlated well with the lattice-matching of $NbSi_2$ to Si(111).

The only stoichiometric silicide in the Mg-Si system, Mg_2Si, has an anti-CaF_2 structure (Fm3m) with a lattice constant of 0.635nm, where the silicon atoms occupy face-centerd cubic sites and the 8 magnesium atoms are located at $\pm(a/4, a/4, a/4)$ points in the unit cell, where a is the lattice constant. Looking for good-quality epitaxial layers, it was noted[129] that the lattice mismatch between the silicide and the silicon substrate was less than 3%. The Mg_2Si lattice parameter of 0.635nm means that a side of the unit cell on the $Mg_2Si(111)$-(1x1) surface is equal to 0.449nm. This in turn means that it is about 1.3% longer than the 0.4434nm that is equal to $(2/3)\sqrt{3}$ times the surface lattice parameter (0.384nm) of the ideal Si(111)-(1x1) surface. Due to this small mismatch, the high-quality epitaxial growth of Mg_2Si onto monocrystalline silicon was assured.

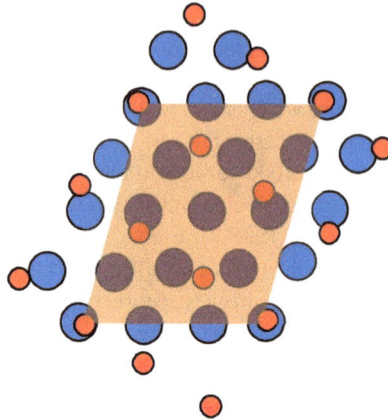

Figure 18. Interfacial matching B between NbSi$_2$ and silicon
Red: niobium or silicon in NbSi$_2$, blue: silicon in substrate

When magnesium films of various thickness were deposited[130] onto (111) silicon substrates by magnetron sputtering and annealed in argon, the resultant Mg$_2$Si thin films were polycrystalline and had a preferred Mg$_2$Si(220) orientation regardless of the original magnesium film thickness or annealing temperature. X-ray diffraction indicated that high-quality silicide films were produced by annealing at 400C for 5h. Films which were annealed below 350C or at above 450C contained magnesium crystallites or magnesium oxide. The grains in the polycrystalline films were 1 to 5μm in size, and the texture of the silicide films became denser and more homogeneous as the thickness of the magnesium film increased. There were 2 main types of epitaxial relationship between Mg$_2$Si(220) and silicon on the (111) substrate (figure 19). The first type was Mg$_2$Si(220)‖Si(111); Mg$_2$Si[001]‖Si<11$\bar{2}$> where the mismatches were Si{111}‖Mg$_2$Si{220} = -4.5% and (5xSi{111})‖(4xMg$_2$Si{220}) = -6.4%. The other type was Mg$_2$Si(220)‖Si(111); Mg$_2$Si[001]‖Si<$\bar{1}$10> where the mismatches were (5xSi{111})‖(7xMg$_2$Si{220}) = -5.5% and (3xSi{111})‖(2xMg$_2$Si{220}) = -10.3%. There were also higher-order mismatches between other Mg$_2$Si planes and silicon (111), such as Mg$_2$Si(111), Mg$_2$Si(200) and Mg$_2$Si(311).

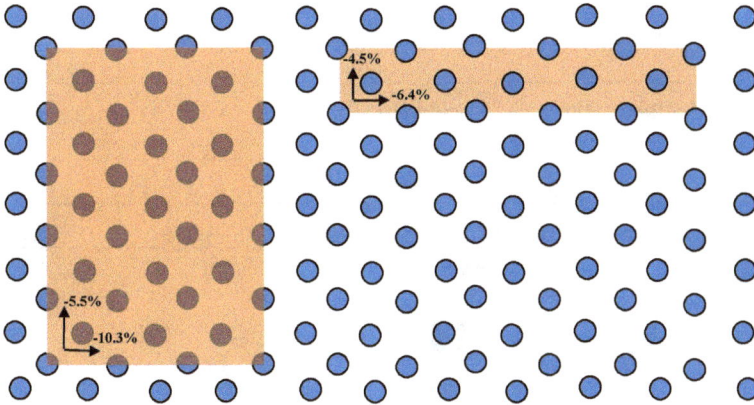

Figure 19. Epitaxial relationships between Mg₂Si and silicon (111)

An amorphous interlayer formed during the solid-state diffusion of ultra-high vacuum deposited polycrystalline gadolinium thin films on (001) silicon surfaces[131]. Following the growth of the amorphous interlayer, epitaxial hexagonal $GdSi_{2-x}$ formed at the amorphous interlayer interface with (001)Si at 225C. A relatively small lattice mismatch between $GdSi_{2-x}$ and (001)Si, as compared with that for rare-earth silicides (table 5), aided the epitaxial growth of $GdSi_{2-x}$ on (001)Si and thus impeded further growth of the amorphous interlayer. It is possible to create a parallel array of nanowires by growing thin layers of rare-earth silicides on a monodomain vicinal (001) silicon surface. The growth of compact silicide islands which usually co-exist with nanowires first has to be suppressed[132]. Nanowire growth can then be optimized by manipulating the growth kinetics or choosing a suitable rare-earth metal. Gadolinium tends to create nanowires under a wider range of conditions than do other metals, and some of the differences in behaviour of the metals are attributed to the effect of biaxial strain.

Table 5. Lattice mismatch of rare-earth silicides on silicon

Silicide	Matching Planes	Epitaxial Condition	Mismatch a (%)	Mismatch b (%)
$GdSi_{2-x}$	$(10\bullet0)GdSi_{2-x}\|(001)Si$	$[00\bullet1]\|[1\bar{1}0]$	0.89	-0.41
$TbSi_{2-x}$	$(10\bullet0)TbSi_{2-x}\|(001)Si$	$[00\bullet1]\|[1\bar{1}0]$	0.18	-1.04
$ErSi_{2-x}$	$(10\bullet0)ErSi_{2-x}\|(001)Si$	$[00\bullet1]\|[1\bar{1}0]$	-1.1	-2.41

a and b are the sides of the common unit cell

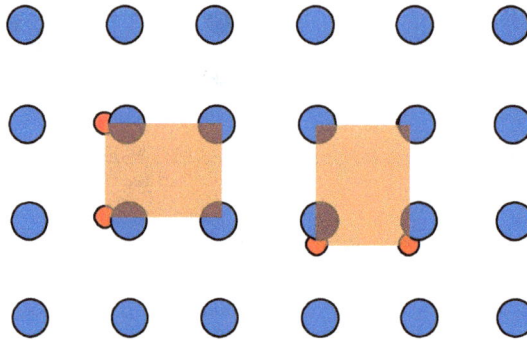

Figure 20. Two forms of $ErSi_{2-x}$ epitaxial growth on the (100) silicon plane
Blue: silicon, red: $ErSi_{2-x}$

Epitaxial erbium silicide, $ErSi_{2-x}$, grew on a (100) silicon substrate following deposition[133] of an erbium film and annealing at above 400C. The films consisted of 2 types of domain having 2 different azimuthal orientations which were arranged at 90° to one another (figure 20)[134]. The epitaxial relationships between the hexagonal film and the (100) silicon substrate were: $[00\bullet1]ErSi_{2-x}\|[01\bar{1}]Si$ and $[00\bullet1]ErSi_{2-x}\|[0\bar{1}\bar{1}]Si$ in the $(1\bar{1}\bullet0)ErSi_{2-x}\|(100)Si$ plane relationship. The two types of domain in the $ErSi_{2-x}$ film were equivalent in volume fraction and crystalline quality. Erbium-silicide thin films were grown[135] on (100) silicon substrates under high vacuum by depositing erbium alone, or by co-depositing erbium and silicon, followed by annealing (800 to 870C, 0.5h). They were between 35 and 50nm thick and consisted of patches of up to a few hundred nm in extent. After depositing erbium alone, epitaxial growth of a tetragonal phase with a single epitaxial mode occurred and was assumed to be induced by the substrate orientation. In the case of co-deposited samples, this same epitaxy occurred together with so-called multiple-variant epitaxial patches having an hexagonal ErSi$_2$ structure.

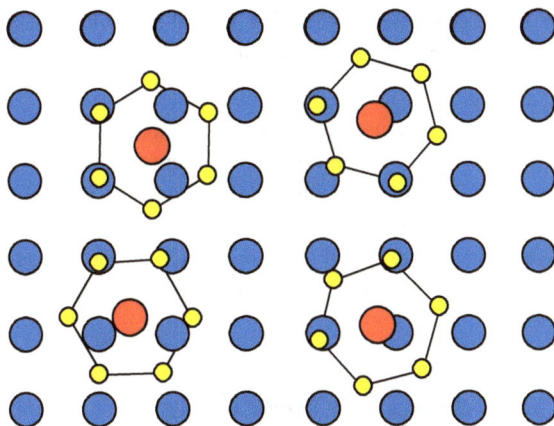

Figure 21. In-plane orientations of epitaxial HfB$_2$ film on silicon substrates
Silicon: blue, hafnium: red, hydrogen: yellow
Clockwise from top-left: (00•1)HfB$_2$||(001)Si; [11•0]HfB$_2$||[110]Si,
(00•1)HfB$_2$||(001)Si; [11•0]HfB$_2$||[100]Si, (00•1)HfB$_2$||(111)Si;
[11•0]HfB$_2$||[110]Si, (00•1)HfB$_2$||(111)Si; [1$\bar{1}$•0]HfB$_2$||[110]Si

Chemical vapour deposition has been used[136] to produce epitaxial HfB$_2$ films on silicon substrates at 1000C, by etching through a pre-existing SiO$_2$ layer. In spite of their differing symmetries, the films are oriented such that: (00•1)HfB$_2$||(001)Si. The X-ray rocking curves are very sharp, with a full-width at half-maximum of 0.076°. Plan-view scanning electron microscopy shows that the film consists of micrometer-sized domains on the substrate, while X-ray pole figures and electron back-scattering diffraction reveal the existence of 4 types of domain (figure 21), with in-plane orientations which are rotated by 45° with respect to one another: [2$\bar{1}$•0]HfB$_2$||[100]Si, [2$\bar{1}$•0]HfB$_2$||[110]Si.

As-grown silver clusters on hydrogen-terminated (111) silicon surfaces were characterized[137] by means of high-resolution electron imaging and diffraction, showing that they grew along [111], normal to the substrate. At a critical average diameter of about 12nm, the clusters became epitaxially oriented: [110]Ag||[110]Si. At this orientation, a silver cluster had a buried interface of 4 x 4 coincidence lattice with silicon and was strained to the extent of -0.32%. Epitaxy coincided with a transition from initial droplet growth to the fractal coalescence of small silver clusters, thus suggesting that epitaxy developed during coalescence. By using an extended rigid-interface model it was shown that, for a highly mismatched lattice, a well-developed interface-energy minimum was possible only for large clusters comprising several coincidence-lattice cells (figure 22). The gain in interface energy for large clusters was the driving force, and cluster-coalescence was the mechanism of cluster reorientation.

Aluminium films, formed[138] by ion-cluster beam-deposition, and silver films, formed by molecular beam epitaxy, on (111) silicon substrates were studied using ion-channelling techniques[139]. The aluminium and silver films were epitaxial, in spite of their large (circa 25%) lattice-mismatches with respect to the silicon substrate. Both films contained dislocation loops, apparently due to the large mismatch. There existed so-called semi-coherent interfaces in which 4 aluminium lattice-planes almost perfectly matched 3 silicon lattice-planes.

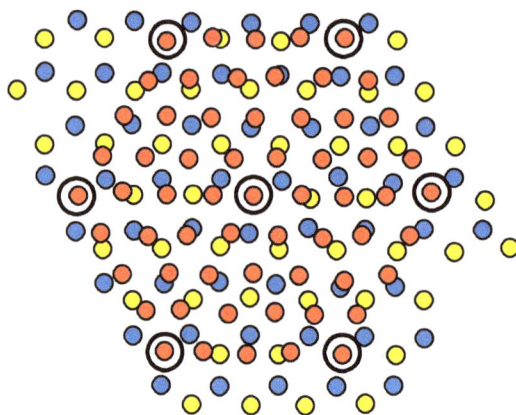

Figure 22. Geometrical model for the coincidence lattice of silver (red) on silicon (blue) hydrogen atoms: yellow. Heavy circles indicate coincidence sites

Plan-view micrographs showed[140] that silver was epitaxial on both the (111) and (100) planes of silicon when deposited at room temperature and annealed at up to 400C. Cross-sectional lattice images showed that 2 types of interface were possible for the Ag‖(111)Si interface; one being related to the other by a 180° rotation of silver around the [111]

Materials Research Forum LLC
https://doi.org/10.21741/9781644900475

silver axis. One of them, A, corresponded to $(111)Si\|(111)Ag$, with $[110]Si\|[110]Ag$. The other, B, corresponded to $(111)Si\|(111)Ag$, with $[114]Si\|[110]Ag$. The interface was partially coherent, with one misfit dislocation every per 4 silver lattice planes. There appeared to be no amorphous or disordered layer at the interface. Lattice images of the $Ag\|(100)Si$ interface revealed 2 epitaxial orientations: $(111)Ag\|(100)Si$, with $[011]Ag\|[011]Si$ and $(100)Ag\|(100)Si$, with $[011]Ag\|Si[011]Si$.

Islands of silver, grown on silicon at room temperature, have been shown[141] to have highly-preferred heights and flat tops. At coverages of about 1 monolayer, the islands are nearly all of (111) orientation and are 2 atomic layers thick. Annealed films which are of the order of 40 monolayers thick yield 2 closely-spaced Ag(111) diffraction-peaks. One of these is weak and broad, while the other is narrow and more intense. The broad peak corresponds to an average expansion of 0.21%, while the other peak is due to a contraction by 0.17%, of the Ag(111) planar spacing as compared to the bulk value. The coexistence of compressive and tensile strains is attributed to temperature changes and to a thermal expansion coefficient mismatch between silver and silicon. This suggests that epitaxial systems – if only incomplete ones - can sometimes accommodate incompatibilities without necessarily requiring common matching lattices or the creation of defects.

The room-temperature growth of 1000 to 1500Å silver films on silicon substrates was such that films always contained a mixture of epitaxial grains and randomly-oriented (111) grains[142]. These orientations were $(111)Ag\|(111)Si$, with both type-A ($<110>Ag\|<110>Si$) and type-B ($<110>Ag\|<114>Si$) twins, $(110)Ag\|(110)Si$ with $<001>Ag\|<001>Si$ and $(100)Ag\|(100)Si$ with $<011>Ag\|<011>Si$. All of these orientational relationships matched 3 silicon atomic rows with 4 silver rows. Defined by the ratio of epitaxial to non-epitaxial grains, the degree of epitaxy decreased in the order: (111) - 110) - (100). Increasing the Ag^+ ion-energy during deposition generally decreased the ratio, and the annealing of $Ag\|Si(100)$ films led to preferential (100) grain growth.

When the size- and shape-dependent energetics of silver nanocrystals on a hydrogen-passivated (111) silicon surface were studied using transmission electron microscopy and molecular dynamics simulations, it was found[143] that the equilibrium orientation, as dictated by interface-energy minimization, depended upon the interface size and shape. In the case of interfaces which approached a coincidence site lattice cell in size, fluctuations of a single atom at the interface could theoretically lead to large variations in nanocrystal orientation. Reflection high-energy electron diffraction was used[144] to characterise silver films which were grown on hydrogen-terminated Si(111) and Si(100) substrates by means of molecular beam epitaxy. This revealed the occurrence of (111)-oriented silver growth on (111) silicon substrates at room temperature or 275C, with $[011]Ag\|[011]Si$.

Scanning electron microscopy revealed islands of the order of 3000Å, for samples grown at 275C, while predominant (100)-oriented growth of silver on Si(100) occurred at room temperature. Silver films which grew on (100) silver substrates at 275C exhibited (100)-oriented epitaxial growth, with [001]Ag∥[001]Si, and island growth with a size of 3000Å.

When aluminium is deposited[145] onto (001) silicon surfaces by sputtering, the orientational relationship is: (110)Al∥(001)Si. Its crystallography has long been analysed from the point of view of lattice-matching geometry and misfit strain-energy. The use of superlattice unit-cells has proved useful in evaluating the lattice correspondence between aluminium and silicon planes at interfaces. Also reported[146] is a (111)Al∥(111)Si orientational epitaxy in which there is a 19° rotation with respect to the parallel epitaxy. This could again be explained by using a geometrical lattice-matching concept.

The morphology of (111) aluminium films, grown[147] on the (111)-(7x7) silicon surface at ambient temperatures or 145K, exhibited – in the latter case - a well-defined critical thickness (4 monolayers) at which atomically-flat aluminium films of marked stability formed[148]. Formation of the flat films at this critical thickness completed the Al/Si(111) interface and led to subsequent homo-epitaxial style layer-by-layer growth.

When aluminium layers were deposited[149] onto the (100) face of silicon at room temperature, they grew epitaxially on Si(100)-(2 x 1) with the familiar orientational relationship: Al(110)∥(100)Si. The aluminium layer comprised 2 types of (110)-oriented domain which were rotated by 90° with respect to each other, according to the relationships: Al[001]∥Si[011] or Si[01$\bar{1}$]. There was a continuous change from the original Si(100)-(2 x 1) pattern to a (1 x 1) pattern following the deposition of 2 monolayers. On an off-oriented and almost single-domain Si(100)-(2 x 1) surface, essentially single-domain monocrystalline Al(110) layers were obtained, with just a few 90°-rotated crystals. The aluminium domains were attributed to the Si(100)-(2 x 1) surface domains.

Table 6. Epitaxial relationships of various metals on silicon

Metal	Temperature (C)	Face	Epitaxy	Q	R
Ag	150	(111)	{111}	1	CC
Ag	280	(111)	{111}	1	CC
Ag	400	(111)	{111}	1	CC

Ag	150	(100)	{100}	1	CC
Ag	280	(100)	{100}	1	CC
Ag	400	(100)	{100}	1	CC
Ag	280	(110)	{110}	n	none
Al	200	(111)	{100}	3	‖
Al	250	(111)	{100}	3	‖
Al	280	(111)	{100}	3	‖
Al	325	(111)	{111}	1	CC|
Al	400	(111)	{111}	1	CC
Al	200	(100)	{110}	2	‖
Al	250	(100)	{110}	2	‖
Al	280	(100)	{110}	2	‖
Al	350	(100)	{110}	2	‖
Al	400	(100)	{110}	2	‖
Al	280	(110)	{110}	n	none
Au	200	(111)	{hkl}	n	none
Au	220	(111)	{hkl}	n	none
Au	200	(100)	{110}	2	‖
Au	220	(110)	{hkl}	n	none
Co	280	(111)	{hkl}	n	none
Co	300	(100)	{hkl}	n	none
Co	280	(110)	{hkl}	n	none
Cr	280	(111)	{hkl}	n	none
Cr	280	(100)	{hkl}	n	none
Fe	280	(111)	{111}	1	CC
Fe	280	(100)	{hkl}	n	none
Fe	280	(110)	{hkl}	n	none

Ni	280	(111)	{110}	3	‖30°
Ni	350	(111)	{100}	3	‖30°
Ni	280	(100)	{100}	1	CC
Ni	280	(110)	{hkl}	n	none
Ti	280	(111)	{hkl}	n	none
Ti	350	(100)	{hkl}	n	none
Ti	280	(110)	{hkl}	n	none
V	300	(111)	{hkl}	n	none
V	300	(100)	{hkl}	n	none
V	280	(110)	{hkl}	n	none

‖: parallel close-packed directions, ‖30°: close-packed directions enclose an
angle of 30°, CC: cube-on-cube, R: orientational relationship, Q: number of
orientation variants

Epitaxial face-centerd cubic, body-centerd cubic and hexagonal close-packed metal films
were deposited[150] onto the (111), (110) and (100) surfaces of silicon and germanium at
various temperatures. Simple epitaxial relationships were found mainly for those face-
centerd cubic metals which formed binary eutectics with the substrates. Silver was
unusual in having a monocrystalline cube-on-cube relationship with all 6 substrates
(tables 6 and 7). Aluminium and gold formed bi-crystalline films on (100) substrates, but
differing relationships on (111) substrates. Silicide formers such as copper and nickel,
and all of the body-centerd cubic and hexagonal close-packed metals, did not have
epitaxial relationships with respect to most of the substrates. Epitaxial single-crystalline,
bi-crystalline and tri-crystalline films of several metals could be grown by using a silver
buffer layer. It was concluded that an epitaxial film can be easily grown if the film forms
a simple eutectic or monotectic with the substrate. The exact epitaxial relationships
thereafter depend upon crystallographic factors governing epitaxy and upon the substrate
structure[151].

Layers of Ni_2Ge and NiGe were found to grow epitaxially on the (111) plane of
germanium at 160 and 250C, respectively. Epitaxial Ni_2Ge regions, some 30nm in size,
covered nearly all of the surface area. The orientational relationships were:
[00•1]Ni_2Ge‖[111]Ge and (10•0)Ni_2Ge‖($\bar{2}$20)Ge. The best NiGe epitaxy was observed
for samples which had been annealed at 500C for 1h. The orientational relationships

were: $[010]\text{NiGe}\|[111]\text{Ge}$ and $(002)\text{NiGe}\|(\overline{2}20)\text{Ge}$. The average size of the epitaxial regions was about 4.5μm. The areal fraction of epitaxial regions of NiGe was about 70%.

Table 7. Epitaxial relationships of various metals on germanium

Metal	Temperature (C)	Face	Epitaxy	Q	R
Ag	150	(111)	{111}	1	CC
Ag	280	(111)	{111}	1	CC
Ag	400	(111)	{111}	1	CC
Ag	150	(100)	{100}	1	CC
Ag	280	(100)	{100}	1	CC
Ag	400	(100)	{100}	1	CC
Ag	280	(110)	{110}	1	CC
Al	216	(111)	{100}	3	‖
Al	280	(111)	{100},{110}	n	none
Al	325	(111)	{100},{111}	n	none
Al	400	(111)	{111}	1	CC‖
Al	280	(100)	{110},{100}	4	‖,CC,CC45°
Al	400	(100)	{110}	2	‖
Al	280	(100)	{hkl}	n	none
Au	200	(111)	{111}	2	CC,CT
Au	280	(111)	{111}	2	CC,CT
Au	200	(100)	{100}	2	‖
Au	220	(100)	{110}	2	‖
Au	280	(100)	{110}	2	‖
Au	200	(110)	{hkl}	n	none
Co	280	(111)	{hkl}	n	none
Co	280	(100)	{hkl}	n	none

Co	280	(110)	{hkl}	n	none
Cr	280	(111)	{hkl}	n	none
Cr	280	(100)	{hkl}	n	none
Cr	280	(110)	{hkl}	n	none
Fe	280	(111)	{hkl}	n	none
Fe	280	(100)	{100}	1	CC
Fe	280	(110)	{hkl}	n	none
Ni	280	(111)	{111}	1	CC
Ni	280	(100)	{100}	1	CC45°
Ni	280	(110)	{110}	n	none
Ti	280	(111)	{hkl}	n	none
Ti	280	(100)	{hkl}	n	none
Ti	280	(110)	{hkl}	n	none
V	280	(111)	{hkl}	n	none
V	280	(100)	{hkl}	n	none
V	280	(110)	{hkl}	n	none

‖: parallel close-packed directions, CT: cube-twin relationship, CC45°: foil normal is a common 100 direction and crystals are misoriented by 45° about this common direction, CC: cube-on-cube, R: orientational relationship, Q: number of orientation variants

Substrates of the form, Ge/Si(111), were first prepared[152] by depositing a monolayer of germanium onto silicon (111)-(7x7) surfaces and, following deposition, the reflection high-energy electron diffraction pattern was found to have changed to (1x1). Germanium and silver deposition was then carried out at 550C. The deposition of silver on Ge/Si(111) substrates, up to a depth of 10 monolayers, produced a prominent ($\sqrt{3}$ x $\sqrt{3}$)-R30° reflection high-energy electron diffraction pattern, together with a streak structure arising from the (111) silver surface. Scanning electron microscopy revealed the formation of silver islands plus a large fraction of open area which was assumed to have the silver-induced ($\sqrt{3}$ x $\sqrt{3}$)-R30° structure on the Ge/Si(111) surface. X-ray diffraction detected the presence of only a (111) silver peak; thus indicating the epitaxial growth of silver on the Ge/Si(111) surfaces. The use of antimony as a surfactant[153] enables the growth of arbitrarily thick continuous smooth defect-free germanium films on (111) silicon substrates. The lattice mismatch (4.2%) is compensated for by a periodic network of interface dislocations. The strain fields of these dislocations give rise to an elastic surface deformation of less than 0.1nm. The dislocations do not intersect at a single point, but instead form an extended 1.8nm node. The lateral lattice constant fluctuates by about ±1% around the bulk germanium lattice constant, but the lattice constant of the germanium film is well-defined only for thicknesses greater than about 10nm. With increasing film thickness, the dislocation strain fields increasingly interact and result in a greater regularity of the dislocation spacings.

Lattice-mismatch studies of $Si_{0.85}Ge_{0.15}$ epitaxial layers, grown by chemical vapor deposition at 1100C onto (111)-oriented silicon substrates, performed using weak-beam dark-field transmission electron microscopy, revealed[154,155] the existence of a regular misfit dislocation network. This resembled the honeycomb network of edge-type dislocations which was predicted by 0-lattice theory. Contrary to theory, the misfit dislocations were dissociated into misfit partials. Triangular planar faults bounded by misfit partials were also observed, even though 0-lattice theory predicted dislocation nodes. High-resolution electron microscopy showed that the planar defects were intrinsic and extrinsic stacking faults, and that their regular arrangements generated 3 different atomistic interface structures. It was concluded that the formation of interfacial stacking-faults, via the dissociation of misfit-dislocation nodes, minimized the energy of the (111)GeSi/Si interface.

The interfacial misfit between a zinc oxide film and (100), (110) or (111) silicon substrates, when examined using 0-lattice theory, was found[156] to be such that the lattice sites of ZnO and silicon at the (00•2)ZnO∥(111)Si interface were in one-to-one correspondence. The misfit at the other interfaces was relatively large. The dislocation densities at the interfaces were deduced from the dislocation networks and were found to be 1.638, 0.588 and 0.718/nm, respectively, for the (111), (110) and (100) silicon substrates. Upon considering the lattice correspondence and the dislocation density, the degree of interface misfit was seen to decrease in the order: (100), (110), (111).

An investigation was made[157] of the epitaxial growth of HfN films, and the sequential mono-oriented growth of Al/HfN bilayer films on the (001) or (111) planes of silicon. Stoichiometric HfN was prepared on silicon via reactive sputtering, and an aluminium film was then sputter-deposited onto the HfN/Si layers. The HfN films which were deposited on silicon grew epitaxially, with the orientational relationships: (001)[110]HfN∥Si(001)[110]Si and (111)[002]HfN∥(111)[002]Si.

When TiN films were sputtered[158] onto (100) silicon surfaces at 400C under flowing nitrogen the films were stoichiometric, with a single oriented (200) plane, and had the same lattice constant as that of the bulk material. The insertion of a stoichiometric $Al_3Ti(112)$ interlayer produced a structure of the form: $(111)Al\|(112)Al_3Ti\|(200)TiN\|(100)Si$, having a small lattice-mismatch at each interface.

The coherent growth of α-Si_3N_4 precipitates was observed[159] in a silicon matrix following the implantation of 150keV N^+ into (110) silicon. The near-channelling conditions produced a band of discrete precipitates at about 0.5µm below a continuous polycrystalline buried nitride layer. No misfit dislocations or strain contrast were observed in the silicon matrix in spite of a 10% lattice mismatch along the Si_3N_4 [00•1] direction and a 1% mismatch along directions perpendicular to [00•1]. The mismatch seemed to be accommodated within the precipitate by a mosaic of monocrystalline sub-units which were coherent with silicon at the $(00•1)Si_3N_4\|\{111\}Si$ interface, but were incoherent on perpendicular planes.

The possibility of growing GaN films on planar high-index silicon (hhk) substrates, in a non-polar orientation, has been examined theoretically[160]. Possible substrates were short-listed by requiring that they be well lattice-matched, over a length-scale of several unit cells, to GaN in the non-polar m-plane orientation. The selected orientations were then used to construct atomistic models of the GaN/Si(hhk) interface. Density functional theory was used to calculate the formation energy of each non-polar interface and to compare it to those of the competing polar interface. It was concluded that Si(112) and Si(113) were potentially favorable substrates for the growth of non-polar m-plane GaN.

Although there is a large lattice mismatch between bismuth and silicon, it is possible to grow atomically smooth and essentially defect-free epitaxial (111) bismuth films on (001) silicon substrates. The remaining mismatch (2.3%) is accommodated by the formation of interface-confined periodic arrays of edge-type dislocations. A comparison of the resultant surface distortions with elasticity theory indicated[161] that a Burgers vector of ½[110] was associated with the misfit-dislocation array. The bismuth film relaxed, towards its bulk lattice constant value, with increasing thickness.

Lattice-Matching to Other Semiconductors

Beginning with the concept of the covalent radius, simple rules were long ago offered for the lattice-matching of tetrahedrally-bonded semiconductors. It had been noted that the lattice constants of semiconducting compounds which were made up of elements from a given row of the periodic table were essentially the same. This suggested that the tetrahedral radii of atoms in a given row should also be approximately constant. The values of the tetrahedral radius are exactly equal for elements in the second and third rows (table 8), but those in the first row are markedly affected by core electronic states.

The ability to deduce the bond-length simply by adding the tetrahedral radii of the component elements, plus the relationship of the bond-length to the lattice parameter, then makes it easy to choose pairs of semiconductors having well-matched lattice parameters; at least in a one-dimensional sense. For the second and third rows of the table, the chemical formula of a binary semiconductor compound can be written as A_iB_j, where A is a group-II, III or IV element, B is a group-VI, V or IV element and i and j are row-numbers. Closely lattice-matched semiconductor pairs, $A_iB_j/A'_mB'_n$, are then those for which i = m and j = n, or for which i = n and j = m. In particular, semiconductors involving elements from the same row in the table obey the rule, given that i = j = m = n.

Table 8. Tetrahedral radii of group-II, III, IV, V and VI elements

II	III	IV	V	VI
	Al (1.230Å)	Si (1.173Å)	P (1.128Å)	S (1.127Å)
Zn (1.225Å)	Ga (1.225Å)	Ge (1.225Å)	As (1.225Å)	Se (1.225Å)
Cd (1.405Å)	In (1.405Å)	Sn (1.405Å)	Sb (1.405Å)	Te (1.405Å)

The experimentally determined bond-lengths of semiconductor pairs from rows 2 and 3 of the table did indeed exhibit only small discrepancies between their lattice parameters. The method could be modified so as to treat semiconductors involving row-1 elements, or mercury compounds. The covalent radius of mercury could be deduced from $r_{Hg} = d_{HgB} - r_B$, where d_{HgB} was the experimentally known nearest-neighbor distance in a mercury compound involving Te, Se or S, and r_B was the corresponding tetrahedral radius, of the Te, Se or S atom in the table. This indicated that $r_{Hg} = 1.402$Å; essentially the same value as r_{Cd}.

The overall rule for lattice-matching all II-VI, III-V and group-IV semiconductors was then that any combination of cations and anions which was symmetrical about column-IV led to an acceptable match. This rule was shown to identify the 36 well-matched combinations (table 9) among 210 possible pairs, with no mismatch being worse than 1.5% and most being better than 0.5%. The good matches included II-VI and III-V compounds, group-IV elements and manganese chalcogenides. The list of well-matched pairs of semiconductors which was obtained in that way included cases which were of interest in the context of hetero-epitaxy. The covalent radius-based procedure could be further extended[162] so as to cover ternary alloys which might be used to obtain perfect lattice-matching or to tailor the degree of mismatch. Also considered was the possibility of anchoring monocrystalline films of cubic phases of iron, cobalt, nickel and manganese to tetrahedrally-bonded semiconductor substrates.

Table 9. Experimental bond-lengths and relative lattice-mismatches of semiconductor pairs

AB‖A'B'	$d_{AB}(\text{Å})$	$d_{A'B'}(\text{Å})$	$2\lvert(d_{AB}-d_{A'B'})\rvert/(d_{AB}+d_{A'B'})$ (%)
CdTe‖InSb	2.806	2.805	0.04
AlAs‖Ge	2.451	2.450	0.04
GaAs‖Ge	2.448	2.450	0.08
ZnTe‖GaSb	2.641	2.639	0.08
CdSe‖InAs	2.621	2.623	0.08
AlAs‖ZnSe	2.451	2.454	0.12
AlAs‖GaAs	2.451	2.448	0.12
CdTe‖Sn	2.806	2.810	0.14
ZnSe‖Ge	2.454	2.450	0.16
InSb‖Sn	2.805	2.810	0.18
HgSe‖GaSb	2.634	2.639	0.19
ZnSe‖GaAs	2.454	2.448	0.25
HgTe‖InSb	2.798	2.805	0.25
HgSe‖ZnTe	2.634	2.641	0.27
HgS‖InP	2.534	2.541	0.28
HgTe‖CdTe	2.798	2.806	0.29
AlP‖GaP	2.367	2.360	0.30
GaP‖Si	2.360	2.352	0.34
HgSe‖InAs	2.634	2.623	0.42

HgTe‖Sn	2.798	2.810	0.43
ZnS‖Si	2.342	2.352	0.43
CdS‖HgS	2.522	2.534	0.47
HgSe‖CdSe	2.634	2.621	0.49
AlSb‖ZnTe	2.657	2.641	0.60
GaSb‖InAs	2.639	2.623	0.61
AlP‖Si	2.367	2.352	0.64
ZnTe‖InAs	2.641	2.623	0.68
CdSe‖GaSb	2.621	2.639	0.68
AlSb‖GaSb	2.657	2.639	0.68
CdS‖InP	2.522	2.541	0.75
ZnTe‖CdSe	2.641	2.621	0.76
GaP‖ZnS	2.360	2.342	0.77
AlSb‖HgSe	2.657	2.634	0.87
AlP‖ZnS	2.367	2.342	1.06
AlSb‖InAs	2.657	2.623	1.29
AlSb‖CdSe	2.657	2.621	1.36

There was good agreement between calculated and experimental data for AlAs/GaAs and $Al_xGa_{1-x}As$/GaAs. That for $Ga_xIn_{1-x}As$/GaAs heterostructures was inferior due to the failure of pseudomorphism for very thick epitaxial layers and because of appreciable misfit. In the GaP/Si system, smaller misorientations were found for substrate deviations of 0 to 10°. The failure of pseudomorphism here was attributed to the difference in the chemical-bond types of the adjacent crystals.

It was shown[163] that *in situ* electron diffraction measurements permitted $In_{0.53}Ga_{0.47}As$ to be lattice-matched to InP during growth. This involved analysis of the dynamics of intensity oscillations during electron diffraction in order to detect when lattice-matched growth was occurring. This method could closely lattice-match InGaAs to InP, GaInP to GaAs and InAlGaAs to InP. It was noted[164] that the HCl, flowing over a gallium boat, which was required to maintain a lattice-matched composition of InGaAs against InP under equilibrium conditions, was particularly important. Increasing the amount of unreacted HCl markedly decreased the HCl(Ga) fraction that was necessary for lattice-matching. This sensitivity could explain why the lattice-matching HCl(Ga) fraction for different systems might be very different for apparently similar growth conditions. It was also in such systems that so-called automatic lattice-matching molecular beam epitaxial growth was first reported[165]. This effect was noted in the growth of AlInAs on InP at a substrate temperature which was sufficiently high for appreciable indium desorption to occur and was attributed to the effect of strain upon the growth mechanism, although the influence of strain upon the thermodynamic equilibrium did not alone appear to be adequate to explain fully the observed effect.

Table 10. Perfect matches between NbTiN and InAsSb or GaInAs surfaces

Semiconductor	Face	a_{alloy}/a_{metal}	Number of cells
InAsSb	$[1\bar{1}0]$	1.414	2
InAsSb	$[1\bar{1}0]$	1.5	9
InAsSb	$[111]$	1.363	13
InAsSb	$[111]$	1.453	19
InAsSb	$[111]$	1.5	9
GaInAs	$[1\bar{1}0]$	1.354	11
GaInAs	$[111]$	1.271	21
GaInAs	$[111]$	1.309	12
GaInAs	$[111]$	1.323	14
GaInAs	$[111]$	1.333	16
GaInAs	$[111]$	1.363	13
GaInAs	$[11\bar{2}]$	1.265	8
GaInAs	$[11\bar{2}]$	1.291	5
GaInAs	$[11\bar{2}]$	1.323	7

The misfit in wafer-fused InP/GaAs interfaces can be accommodated[166] either by an amorphous intermediate layer containing oxygen and indium, or by dislocations. In the latter case, and when no twist is present, the 3.7% lattice mismatch is relaxed by a regular square network of dislocations of pure edge type. When an additional twist is present, a square network of dislocations again results, but those dislocations are then of mixed type. In this case, the geometry of the dislocation network could be predicted by using 0-lattice theory.

In the case of InAsSb [1$\bar{1}$0], matched to vanadium, the match went from a perfect [110] match to a low-strain [113] match as the lattice constant was varied from 6.06 to 6.08Å. Several zero-strain matches were found between InAsSb [1$\bar{1}$0] and the metal. Nickel, copper, lead and vanadium gave perfect matches which were combined with very small unit cells, promising great stability. Many zero-strain solutions were possible, especially in the case of [111] surfaces, and perfect matches were distributed over a range of lattice constants.

The method has also been used to match semiconductors to the superconductor, $Nb_{1-x}Ti_xN$ by varying the scaling parameter (table 10). In the case of InAsSb, two zero-strain matches were found for the [1$\bar{1}$0] surface and one of these was associated with a unit-cell of only 2 alloy-surface unit-cells. Three zero-strain matches were found for the [111] surface, and one was associated with a unit-cell of 9 surface cells.

An epitaxial helical nano-heterostructure between CdS and $ZnIn_2S_4$ has been observed[167], and experimental and theoretical work has shown that mismatches in lattice and dangling bonds between one-dimensional and two-dimensional units control the growth. The well-defined interface produces delocalized interface states.

When layers of ZnTe, ZnSe and ZnCdTe/ZnTe were grown on hexagonal $Zn_{0.05}Cd_{0.95}S(00\bullet1)$ and $CdS_{0.85}Se_{0.15}(11\bullet0)$ by means of molecular beam epitaxy[168], those grown on (11$\bullet$0) substrates had a rough surface while those grown on a (00$\bullet$1) substrate were mirror-smooth, and all of the structures were cubic. In the case of ZnSe-based structures, the <111> and <110> lattice directions of the epilayers coincided with the <00$\bullet$1> and <11$\bullet$0> directions, respectively, of the CdZnS(00$\bullet$1) substrate. A high mismatch led to lattice-relaxation of the epilayers, due to the creation of misfit dislocations. In spite of having a greater mismatch, the ZnTe-based epilayers which were grown on (00$\bullet$1) substrates had a more perfect lattice structure than did ZnSe-based epilayers. The cubic lattice of ZnTe was rotated, by about 15° around the <111> direction, and coincided with the <00$\bullet$1> direction of the (00$\bullet$1) substrate. It was concluded that geometrical lattice-matching occurred during the epitaxial deposition of ZnTe on CdZnS(00$\bullet$1).

X-ray topography provided an accurate guide to the lattice-mismatch between a liquid-phase HgCdTe epitaxial layer and a (111) CdZnTe substrate[169]. A well-defined crosshatch pattern in the 3 <110> directions indicated the existence of a positive room-temperature lattice mismatch. Under conditions of near-perfect (±0.003%) lattice-matching, the crosshatch pattern disappeared, reflecting an absence or near-absence of misfit dislocations near to the interface. A crosshatch-free region occurred for a small (0.02%) positive room-temperature mismatch, and was attributed to differences in the lattice-matching conditions at room temperature and at the growth temperature. Where the film was in tension, giving a negative mismatch, a mosaic pattern was observed rather than crystallographically-oriented crosshatching. The full-width at half-maximum epitaxial-layer rocking curve was minimal in the crosshatch-free zone, with a value that was nearly equal to that of the substrate. The etch-pit density of the HgCdTe layer was very low ($10^4/cm^2$) for perfect room-temperature lattice-matching. In the case of almost lattice-matched layers, crosshatching was present through the thickness of the epitaxial layer, apart from a narrow graded-composition zone located near to the substrate. The crosshatch contrast was attributed to long-range strain fields that were associated with a misfit dislocation network near to the substrate interface. It is interesting to note that etch-pit studies of misfit dislocations in (111)HgCdTe‖CdZnTe heterojunctions, grown by liquid-phase epitaxy, showed[170] that such dislocations were localized at the original surface of the substrate because zinc which diffused into the epilayer during epitaxial growth prevented their movement. In order to ensure lattice-matching between $Hg_{0.7}Cd_{0.3}Te$ and $Cd_{1-y}Zn_yTe$, the required ZnTe mole fraction of $Cd_{1-y}Zn_yTe$ was 2.9%.

Numerous studies have proved that precise lattice-matching is required in order to obtain crystals which possess a high luminescence efficiency. In the case of bromine-doped ZnSSe on GaAs[171], homogeneous scanning electron microscopic cathodoluminescence images were observed for lattice-matched $ZnS_{0.09}Se_{0.91}$/GaAs, while numerous dark spots were observed for lattice-mismatched ZnSe/GaAs. The dislocation density on the (100) surface of the lattice-mismatched system was estimated to be $10^8/cm^2$ according to cross-sectional transmission electron microscopic images, and $10^7/cm^2$ according to etch-pit measurements. The dislocation density decreased to below $10^6/cm^2$ according to transmission electron microscopic images, and to $10^4/cm^2$ according to etch-pit data, for the lattice-matched system. The room-temperature photoluminescence intensity for lattice-matched ZnSSe was also some 10 times greater, than that for lattice-mismatched ZnSe, for a given ($10^{16}/cm^3$) dopant concentration. Lattice-matching was also shown[172] to be an important factor in effective doping and the control of conductivity. When ZnS_xSe_{1-x} epilayers were grown on GaAs at 470C by organometallic vapor-phase epitaxy using dimethylzinc, methylmercaptan and dimethylselenide precursors, and with gallium and nitrogen as donor and acceptor dopants, respectively, it was found that - for x-values of between 0.06 and 0.12 - ZnSSe epilayers with a thickness of about 1μm grew coherently on the GaAs and nitrogen doping was particularly effective here. Within this composition region, gallium-doped epilayers exhibited a low resistivity while those having other sulfur contents exhibited a high resistivity. Defects which were generated during lattice relaxation markedly degraded the electrical properties. Early work[173] on ZnSe/GaAs interfaces at elevated temperatures had shown that, when the stoichiometry of the ZnSe film was not completely maintained at the interface, there was an increase in zinc diffusion into the GaAs with increasing film thickness. When the GaAs substrates were pre-heated, stable ZnSe films could be deposited onto the GaAs. If the ZnSe film was thicker than 7200Å, however, the original n-ZnSe/n-GaAs interface was converted to p-type during annealing (650C, 3h). The fact that interface diffusion depended upon the film thickness was attributed to misfit-induced lattice relaxation occurring at above a critical thickness. This was confirmed by the observation of an extreme thermal stability in the case of a lattice-matched $ZnS_{0.06}Se_{0.94}$/GaAs interface. It was already known[174] that, when ZnS_xSe_{1-x} layers were grown by molecular beam epitaxy using zinc, selenium and ZnS as source materials and with the sulfur mole fraction being varied between x = 0 and x = 0.3, the epilayer quality was improved for those compositions (x = 0.055 to x = 0.069) which lattice-matched the epilayer to the GaAs substrate; as judged by full-width at half-maximum X-ray rocking curve data and photoluminescence spectra. Later work[175] showed that, when ZnSe was grown directly onto a GaAs substrate, 5 electron-traps with activation energies of 0.20, 0.23, 0.25, 0.37 and 0.53eV, were created. The use of a GaAs buffer layer and lattice-matching could reduce the incorporation of the latter 3 traps; implying that they were due to lattice relaxation. The concentration of the first trap was a function of the donor concentration and lattice mismatch, and its appearance was attributed to the existence of a complex of donor-induced defects and dislocations.

X-ray diffraction was used[176] to monitor the hetero-epitaxial growth of indium which formed at the interfaces between thin films of ZnO, and monocrystalline InP substrates. Indium-formation was caused by the thermal degradation of InP during annealing (400 to 700C, 180s), and its progress was controlled by the decomposition of close-packed {111} planes of the phosphide; with the polar nature of those planes playing a crucial role. For all of the orientations studied, (111)A, (111)B, (110) and (100), Indium (101) grew on InP{111} planes and, on those planes, indium crystallites could have 6 possible orientations of the form, (100)In∥(110)InP.

A geometrical model[177] of the interface between a (100) GaAs substrate and a PbTe molecular beam epitaxial layer was used to explain the observation that PbTe layers of 2 orientations, (100) and (111), could form on the GaAs(100) surface. It was shown that one of the best geometrical matches existed for (111)PbTe∥(100)GaAs, with a surface lattice-mesh of 23.984Å x 3.997Å; that is, a surface reconstruction of (6 x 1), giving a substrate surface having only a few nucleation centers. The other best match was: (100)PbTe‖(100)GaAs, with a surface lattice mesh of 7.99Å x 3.997Å; that is, a surface reconstruction of (2 x 1), giving a substrate surface having a large fraction of nucleation centers. The first form occurred in the case of a GaAs surface which was very carefully heat-treated before growth started. The second form occurred when the substrate was heat-treated at a lower temperature, for a shorter time. It is of incidental interest that lattice-matching plays a role in determining the quantum efficiency of heterostructural PbSnTe photodiodes[178]. For instance, a mismatch of less than 2×10^{-4} results in a photodiode having a quantum efficiency of about 43%; very close to the maximum of 50% which is possible without adding an anti-reflection coating. This is to be compared with the value of 28% in the case of a heterojunction involving a mismatch of 4×10^{-3}. The increase in quantum efficiency can be attributed to the high crystal-quality at the hetero-interface and its environs in a lattice-matched system. High defect-densities are found in the mismatched system.

Transmission electron microscopy and photoluminescence spectra revealed[179] differences in the progress of lattice relaxation in (111)CdTe‖(100)GaAs and (133)CdTe‖(211)GaAs heterostructures, where a 14.6% lattice mismatch existed. The lattice of (133)CdTe layers on (211)GaAs substrates rotated, in order to relax misfit strains, around the <011> axis which was perpendicular to the growth direction and parallel to the <011> axis of the substrate. In contrast, the lattice of (111)CdTe layers on (100)GaAs substrates rotated around the <111> axis which was parallel to the growth direction because the glide plane of (111)CdTe was inconsistent with that of (100)GaAs. It was concluded that CdTe layers of high structural quality were more likely to be obtained by using (211)GaAs substrates rather than (100)-oriented substrates. Lattice-matching was studied[180] in (112)CdTe‖(112)GaAs heterostructures, which have a 13.6% mismatch and where intermediate compounds can form at the interface. The tilt-angle of the film lattice, with respect to the substrate lattice, was used to quantify the degree of matching. The film could be tilted when it was grown on a stepped substrate surface having a different interplanar spacing, or when a regular set of misfit dislocations having equal Burgers vectors was present at the interface. Lattice-matching between the molecular beam epitaxial CdTe film and the GaAs substrate was achieved in such a way that the atomic planes of the substrate which formed steps at the substrate surface continued into the film, without producing a set of misfit dislocations having equal Burgers vectors.

In a study[181] of the relationship between lattice-matching and crystalline quality, of GaInNAs films grown on (001) GaAs substrates, the mismatch between the GaInNAs and the GaAs was varied from -0.35 to 0.29%. Coherent growth of the GaInNAs in all of the samples studied was confirmed by (224) X-ray reciprocal space mapping. The full-width at half-maximum of (004) rocking curves went through a minimum at the lattice-matching point, and the full-width at half-maximum did not depend upon the nitrogen content. Raman scattering measurements of a nitrogen-related local-vibrational mode revealed a broad peak at 470/cm for all of the GaInNAs samples. A further clear shoulder peak, at 490/cm, was attributed to a stress-relaxation mode and was observed only for lattice-matched samples. Lattice-matching not only had a notable effect upon crystal-axis orientation but also upon local stress relaxation.

Transmission electron microscopy showed[182] that aluminium, grown on (001) GaAs surfaces by chemical beam epitaxy, formed islands for all deposit thicknesses and exhibited 4 different orientational relationships with respect to the substrate. One of them, (011)Al||(001)GaAs, became dominant during growth and misfit dislocations were observed in the interface, with a Burgers vector of 1/4(110)GaAs. These dislocations were associated with interfacial steps having a height of 1/2[001]GaAs.

When epitaxial layers of MgO were grown[183] on (100) GaAs substrates, films which were deposited at 500C grew with the stoichiometric composition and had a (110) planar orientation. The [1$\bar{1}$0] direction of the MgO(110) plane was parallel to the [01$\bar{1}$] direction of the GaAs(100) plane, with a 4:3 coincident site lattice.

Transmission electron microscopy of hetero-epitaxial films of InSb, grown by metalorganic magnetron sputtering onto the (100) face of GaAs, and moiré-fringe measurements of elastic strains in the InSb films, showed[184] that the dilatational lattice strain decreased with increasing film-thickness. The measured strains were in overall agreement with simple strain-energy criteria for elastic accommodation and coincidence site lattice theory for hetero-epitaxy.

Geometrical and chemical-bonding considerations with respect to the GaAs||CdTe interface predicted[185] that the [211]CdTe direction would grow parallel to the [011]GaAs direction and that: [011]CdTe||[011]GaAs. It was also predicted that the growth direction would be of [111]-type and that an array of lattice defects would be present at the GaAs||CdTe interface, with a period of 2 x 15.9Å along the [011] direction, in order to accommodate the lattice mismatch. These defects could be of both edge and screw type.

The orientational relationships between CoGa and GaAs were deduced[186] to be: [001]CoGa||[001]GaAs and (220)CoGa||(220)GaAs. The Burgers vectors of the interfacial dislocations were ½<101> and ½<011> and were are inclined with respect to the (001) GaAs surface. Nearly all of the CoGa films were epitaxially related to the surface. No interfacial dislocations were observed in most of the epitaxial CoAs films, which were considered to be pseudomorphic with respect to GaAs. The orientational relationships between CoAs and GaAs were: [101]CoAs||[011]GaAs and (020)CoAs||(220)GaAs.

[187]The atomic arrangements of the (100) and (111) GaAs planes were compared for possible matching to the (00•1), (10•0), (20•1), (10•1), (10•2), (12•0), (12•1), (12•2) and (12•4) planes of BeO. The matching was calculated on both a 1:1 and multiple-atom basis for 2 perpendicular directions in the plane of the film. A figure-of-merit was defined which reflected the elastic strain-energy. It was concluded that lattice-matching was of relatively little value for predicting epitaxial relationships in the GaAs/BeO system. On the contrary, it was decidedly misleading because some predicted low-mismatch orientations were not in fact observed, while some of the observed orientations involved a large mismatch.

Experimental studies[188] of molecular beam epitaxial antimony layers on GaSb revealed[189] that, although antimony is rhombohedral and GaSb has a cubic zincblende structure, bi-atomic planar structure of antimony could mimic[190] the (111) plane of the zincblende structure. Near-perfect matching between the zincblende and the rhombohedral lattice could thus be achieved if epitaxy occurred along the (111) direction. Antimony could be grown epitaxially onto both (111)A and (111)B orientations at below 240C, and GaSb could also be grown on antimony epilayers.

In order to determine the conditions which were required in order to ensure exact lattice-matching during the liquid-phase epitaxial growth of GaInAsP layers on InP substrates, an early study[191] was made of the effect of growth-solution composition, solution supercooling, and substrate orientation upon the composition and properties of epitaxial layers which gave rise to laser-emissions in the 1.1 to 1.3µm range. The composition of the GaInAsP epilayer depended markedly upon the growth parameters, rather than upon the lattice mismatch-energy. Multiple liquid-phase epitaxy of $In_{1-x}Ga_xP_{1-y}As_y$/InP double heterojunctions, from a single set of indium-rich melts, was shown[192] to be a useful technique for the study of lattice-matching at heterojunction interfaces and for the growth of large numbers of low-threshold defect-free laser wafers. Liquid-phase epitaxial growth was carried out at 640C, using indium-rich melts which were saturated at 650C, with the aim of determining the effect of lattice mismatch upon the crystal surface morphology; in particular, the misfit dislocation structure resulting from lattice mismatch[193]. The minority carrier (electron) diffusion length in zinc-doped p-$In_{0.80}Ga_{0.20}As_{0.44}P_{0.56}$ (p being about 10^{18}/cm^3) epitaxial layers ranged from 1.7 to 0.76µm and was essentially independent of the degree of lattice mismatch within the range of ±0.2%. The minority carrier diffusion length was controlled mainly by point-defect centers or by direct radiative recombination[194]. Electron traps at 0.60 and 0.67eV below the conduction band were found in undoped n-$In_{0.77}Ga_{0.23}As_{0.35}P_{0.65}$ layers, and the density of the 0.67eV trap was surprisingly dependent upon the lattice mismatch[195].

Lattice-Matching to Sapphire

Symmetry analysis of the test-case of $A^{III}B^{V}$ nitrides on sapphire predicted a number of permitted types of orientational relationship. The observed relationships were generally consistent with the predictions, and the exceptions were attributed to experimental error in the determination of the orientational relationship or to the layers not being monocrystals.

Table 11. Misfit between nitride layers and sapphire substrate

Nitride	Substrate Direction	Substrate Parameter	Layer Parameter	Misfit (%)
GaN	$[2\bar{1}\bullet 0]$	$2a$	$3a$	0.25
GaN	$[\bar{1}2\bullet 0]$	$2a$	$3a$	0.25
GaN	$[1\bar{1}\bullet 0]$	$2a\sqrt{3}$	$3a\sqrt{3}$	0.25
GaN	$[1\bar{1}\bullet 0]$	$a\sqrt{3}$	$\sqrt{(c^2+4a^2)}$	0.56
GaN	$[01\bullet\bar{1}]$	$\sqrt{(c^2/9+a^2/3)}$	c	0.8
AlN	$[2\bar{1}\bullet 0]$	$2a$	$3a$	1.9
AlN	$[1\bar{1}\bullet 0]$	$2a\sqrt{3}$	$3a\sqrt{3}$	1.9
GaN	$[00\bullet 1]$	c	$4a$	2.1
GaN	$[00\bullet 1]$	c	$4a$	2.1
AlN	$[00\bullet 1]$	c	$4a$	2.9
AlN	$[01\bullet\bar{1}]$	$\sqrt{(c^2/9+a^2/3)}$	c	3.0
AlN	$[00\bullet 1]$	c	$4a$	4.3
AlN	$[1\bar{1}\bullet 0]$	$4a\sqrt{3}$	$3\sqrt{(4c^2+3a^2)}$	4.3
AlN	$[01\bullet\bar{1}]$	$\sqrt{(c^2/9+a^2/3)}$	$a\sqrt{3}$	5.0
GaN	$[01\bullet 0]$	$4a\sqrt{3}$	$3\sqrt{(4c^2+3a^2)}$	6.4
GaN	$[1\bar{1}\bullet 0]$	$4a\sqrt{3}$	$3\sqrt{(4c^2+3a^2)}$	6.4
GaN	$[2\bar{1}\bullet 0]$	$\sqrt{(c^2/9+a^2/3)}$	$a\sqrt{3}$	7.1
AlN	$[2\bar{1}\bullet 0]$	$2a$	$\sqrt{(c^2+9a^2)}$	10.6

InN	$[2\bar{1}\bullet0]$	2a	3a	10.8
AlN	$[2\bar{1}\bullet0]$	a	a√3	12.4
AlN	$[1\bar{1}\bullet0]$	a√3	3a	12.4
AlN	$[2\bar{1}\bullet0]$	a	a√3	12.4
GaN	$[2\bar{1}\bullet0]$	2a	√(c²+9a²)	13.1
GaN	$[2\bar{1}\bullet0]$	a	a√3	14.6
GaN	$[11\bullet0]$	a	a√3	14.6
GaN	$[1\bar{1}\bullet0]$	a√3	3a	14.6
GaN	$[2\bar{1}\bullet0]$	a	a√3	14.6
GaN	$[00\bullet1]$	c	2a√3	16.4
AlN	$[00\bullet1]$	c	2a√3	19.6
InN	$[2\bar{1}\bullet0]$	a	a√3	25.0

The orientational relationships of gallium, aluminium and indium nitrides on sapphire substrates were studied[196] theoretically and experimentally (table 11). It was found that the dependence of those relationships upon the orientations of the sapphire and the compound could not be explained in terms of the coincidence site lattice concept, but could be predicted by symmetry analysis.

An attempt was made to clarify whether geometrical matching was essential for epitaxial growth on $(00\bullet1)Al_2O_3$. It was clear from the successful predictions that an interface plane of the layer should also be $(00\bullet1)$. An oddity was that one of the predicted relationships (figure 23) involved an excellent lattice-match, with 3 GaN unit-cells equalling 2 sapphire unit-cells (and also being in agreement with the Curie law), but was not observed experimentally. A different relationship (figure 24) was alone observed, and was also in agreement with the Curie law. This relationship could not be associated with any lattice-match if the sizes of the common unit-cell areas which were taken into account were not very large. It was concluded that, in this system, symmetry analysis furnished much more information than did the criterion of good lattice-matching when predicting the observed orientational relationship.

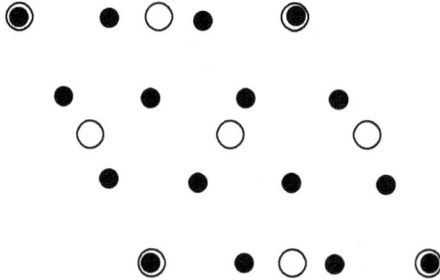

Figure 23. Translation symmetry of the interface between (00•1) GaN and (00•1) sapphire. The nitride translation lattice sites are indicated by small circles and the sapphire translation lattice sites are indicated by large circles. This predicted relationship was not in fact observed.

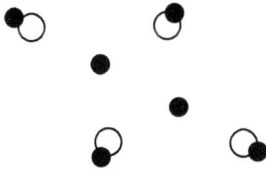

Figure 24. Translation symmetry of the interface between (00•1) GaN and (00•1) sapphire. The nitride translation lattice sites are indicated by small circles and the sapphire translation lattice sites are indicated by large circles. This predicted relationship was observed experimentally.

Sapphire is mainly used as a substrate, for the epitaxial growth of InGaAlN systems, because of a lack of bulk GaN crystals. Early progress was made by using buffer layers and by doping. Cracks in GaN films persist and are attributed to the differing (13.8%) lattice constants and thermal expansion coefficients (-34.2%) of GaN and sapphire. Lattice-matching remains a basic problem in the growth of the high-quality epitaxial films which are required for the preparation of high-performance devices.

In the case of silicon on sapphire, the crystal structures and the lattice constants are very different. Three orientational relationships were predicted, but only one was observed. By means of symmetry analysis, constraints were placed on the orientational relationships. Only the (111) plane of silicon, with 2 non-equivalent azimuth relationships, were predicted to grow on (00•1) sapphire; the lattice-matches being the same in both cases.

A general theoretical approach for treating epitaxial layers having differing cell-parameters and in-plane relaxation anisotropy has been developed[197], together with a covariant description of relaxation in such structures. An iteration method was developed for the evaluation of the parameters on the basis of diffraction data. The accuracy of the approach was proved in the case of ZnO-on-sapphire samples which had been grown at 573 to 1073K, using data measured for various directions of the diffraction plane relative to the sample surface.

The epitaxial orientation of rhodium films with a thickness of 100nm on (11•0) sapphire can be controlled[198] by changing the substrate temperature under low (1.0nm/min) deposition-rate conditions because the lattice mismatches between two-dimensional superlattice cells[199] of (111)Rh‖(11•0)sapphire and (001)Rh‖(11•0)sapphire systems are about the same. The epitaxial orientation of a rhodium film can therefore be changed, from a mixture of (001) and (111) rhodium planes to a single (001) rhodium plane, on (11•0) sapphire, simply by varying the substrate temperature. It was reasoned that, although the closest-packed (111) plane of a face-centerd cubic structure is preferred in the sense of minimum surface energy, under low deposition-rate conditions the sputtered atoms can migrate for relatively long times over the substrate. This was deemed to be equivalent to an increased temperature. On the other hand, lattice-matching leads to a large decrease in film/substrate interfacial energy. Epitaxial growth of (001)Rh could nevertheless be achieved at 1.0nm/min and 500C, even though the film surface energy of (001)Rh was slightly higher than that of (111)Rh. Moreover, the area (7x6 sapphire unit-cells) of the two-dimensional superlattice cell of the (001)Rh‖(11•0)sapphire system was slightly smaller than that (5x11 sapphire unit-cells) of the (111)Rh‖(11•0)sapphire system. In addition, the exact lattice mismatches of the two-dimensional superlattice cell of (001)Rh‖(11•0)sapphire are 0.37% and 0.02%, while the corresponding ones for (111)Rh‖(11•0)sapphire are 0.37% and 0.85%. Because this implies that the interfacial energy of (001)Rh‖(11•0)sapphire is slightly lower than that of (111)Rh‖(11•0)sapphire, epitaxial growth of (001)Rh is favoured on the (11•0) sapphire plane if the substrate temperature is suitable, regardless of the film surface energy. Thus the epitaxial growth of (001)Rh was possible at above 450C because the sputtered rhodium atoms could meander long enough to form a smaller superlattice cell of (001)Rh than of (111)Rh and hence decrease the film/substrate interfacial energy.

Monocrystalline MgB_2 films can be grown[200] onto (00•1) Al_2O_3 substrates by exploiting the perfect lattice-matching ratio, 8:3√3, which exists between the a-axis lattice constants of MgB_2 and Al_2O_3. Selected-area electron diffraction patterns revealed hexagonal MgB_2 films which exhibited a 30° in-plane rotation with respect to the Al_2O_3 substrate.

Helicon-wave excited-plasma sputtering epitaxy of ZnO epilayers showed[201] that the epilayers, when deposited onto (00•1) and (11•0) Al_2O_3 or (00•1) AlN substrates, exhibited c(00•1)-oriented growth. The use of in-plane uniaxially and nearly lattice-matched (11•0) Al_2O_3 substrates permitted the preparation of a-axis locked monodomain ZnO epilayers.

Lattice-Matching to Other Ceramics

A high-resolution transmission electron microscopic study of <110> symmetrical tilt grain boundaries in titanium carbide showed[202] that, at a boundary near to the incoherent (11$\bar{2}$) Σ3 coincidence boundary, there exists a periodic grain-boundary structure which consists of 2 types of atomic arrangement. One was a mirror-symmetrical atomic arrangement in crystals adjacent to the boundary, while the other was a periodic mixed structure of mirror-symmetrical and asymmetrical arrangements. Two types of asymmetrical structure were observed, depending upon slight deviations from the ideal misorientation. This was explained by the displacement shift complete dislocation, b = a/6[11$\bar{2}$], and lattice dislocation, b = a/2[01$\bar{1}$], which extended on their boundaries following dissociation into 2 partial dislocations. Periodic structures, observed in the boundary, could be well-explained in terms of displacement shift complete dislocations, based upon coincidence site lattice theory.

A natural low-angle grain boundary of mixed twist and tilt type, in a magnetite bicrystal, was investigated[203] by means of X-ray diffraction. The reflections which were observed around a 040 matrix reflection agreed well with those predicted by 0-lattice theory. The extra reflections were thought to arise from a largely planar dislocation network. The structural width of the boundary was about 4 unit cells: that is, close to the average spacing of the predicted dislocation network.

In the case of $A^{III}B^V$ materials on spinel, the close-packed planes of the latter are (111), (100) and (110) and the two-dimensional point groups of symmetry are 3M, 4M and MM2, respectively. The elements of the two-dimensional point symmetry of a (001) spinel substrate could not be preserved in the monocrystalline layer, regardless of the crystallographic orientation. Layers could grow on (111) and (110) substrates. Comparison of the theoretical and experimental results showed that polycrystalline growth, with monocrystalline areas, could grow on the (100) spinel plane. When the plane of the substrate deviated from the singular plane by more than 1°, a smooth monocrystalline layer was formed.

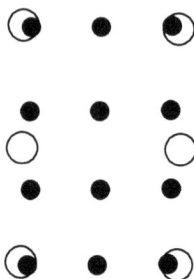

Figure 25. Translation symmetry of the interface between GaP and spinel, (001)GaP||(110)MgAl₂O₄; [110]GaP||[001]MgAl₂O₄. The GaP translation-lattice sites are indicated by small circles and spinel translation lattice are indicated by large circles. No good match was observed.

The orientational relationship of $(111)A^{III}B^{V}/(111)$ spinel corresponded to a good lattice-match at the interface, especially for GaP. When the (110) plane of the spinel was considered, there was no good lattice-match along the $[1\bar{1}0]$ direction (figure 25). If the orientations of the layer and substrate were the same, an exact match would be observed (figure 26). A small lattice misfit was not however an essential condition for single-crystal growth of systems having a given orientational relationship. None of these experimental observations could be explained without considering the symmetry.

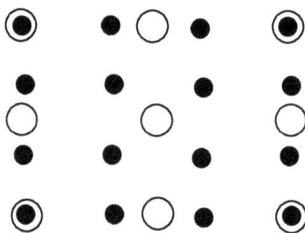

Figure 26. Translation symmetry of the interface between GaP and spinel, (110)GaP||(110)MgAl₂O₄; [001]GaP||[001]MgAl₂O₄. The GaP translation-lattice sites are indicated by small circles and spinel translation lattice are indicated by large circles. A good match was predicted, but not observed.

The optimization of the composition and orientation was considered for the $YBa_2Cu_2O_7/PrGaO_3$ heterostructure. These crystal structures are based upon pseudo-cubic perovskite-type cells and have almost the same lattice parameters but, because they are distorted and ordered differently, their non-primitive Bravais unit-cells have very different parameters. The calculations were therefore based upon reduced pseudo-cubic perovskite cells. Optimization was based upon a scalar cost function. Give that the mismatch between the layer and substrate is small during the initial stages of growth, the layer can sustain it in the form of elastic strain; given that dislocations are ruled out by the above imposed conditions. The elastic energy meanwhile increases with layer thickness up to a critical value, whereupon structural defects appear. Because elastic energy is the driving force for the nucleation of defects, this energy constitutes a suitable cost function, f, defined by $\sqrt{(E_1/E_0)}$, where E_1 is the elastic energy of the epitaxial layer and E_0 is the energy of an idealized cubic/cubic epitaxial pair for which the lattice mismatch is 1% and the thicknesses and elastic moduli of the model and real layers are equal. In the case of cubic heterostructures, this cost function corresponds to the traditional definition of lattice mismatch.

A computational method[204] for predicting optimum substrates for epitaxial growth exploits the use of first-principles calculations of formation energy, elastic strain energy and topological data. In the case of the epitaxial growth of metastable VO_2 on TiO_2, the model involved geometrical unit-cell area matching, between the substrate and the layer, plus consideration of the strain-energy density. There was qualitative agreement with experimental observations. The involvement of a calculated strain-energy, although already implied by previous geometrical models which identify lattice-coincidence with strain, was a useful innovation.

Ferroelectric thin films were grown[205] epitaxially onto a (101)-oriented rutile-structured oxide at 630C, using metalorganic chemical vapor deposition[206]. The long-range lattice-matching between the ferroelectric layer and the bottom rutile layer was such that: $(100)(010)Bi_4Ti_3O_{12}\|(101)TiO_2$. Cross-sectional transmission electron microscopic observations suggested that 7 rutile units corresponded to a single a-/b-axis-oriented $Bi_4Ti_3O_{12}$ unit, and involved a lower misfit-dislocation density, as compared with 8 rutile units being associated with a single $Bi_4Ti_3O_{12}$ unit. This inspired a proposed ion-alignment model in which the titanium layer in the $(101)TiO_2$ structure was supposed to match up with the oxygen layer in the a-/b-axis-oriented $Bi_4Ti_3O_{12}$ film.

Titanium carbide coatings were deposited[207] onto cemented carbide substrates, with and without any interfacial η-carbide (usually Co_6W_6C). In the absence of the latter, the TiC nucleated and grew epitaxially on both the $\{00\bullet1\}$ and $\{10\bullet0\}$ WC surfaces. The epitaxy on the former surface was close-packed plane to close-packed plane: $\{111\}$TiC on

{00•1}WC. The orientational relationship was: (111)TiC||(00•1)WC; [110]TiC||[11•0]WC. In the case of the {10•0}WC surface there were 2 orientational relationships: (001)TiC||(10•0)WC; [110]||[11•0] and (112)TiC||(10•0)WC; [111]||[00•1]WC. The misfit dislocation structures were estimated by using 0-lattice theory. When the η-carbide formed, it replaced WC at the original TiC/WC interface. The associated dissolution of WC occurred preferentially along the {10•0}WC planes.

The concept of a good matching site helps to guide the identification of the coincidence site lattice in two dimensions, and the Burgers vectors in a large misfit system. The good matching site concept was combined[208] with 0-lattice theory in order to calculate the secondary dislocation structure in the habit planes of TiN precipitates in Ni-TiN alloy. Under slight elastic strain, the type-III habit-plane of TiN is predicted to contain a single set of secondary dislocations, as observed experimentally. The type-II habit-plane is predicted to contain 3 sets of secondary dislocations. Two of these are further predicted to relax so as to be almost parallel. The third might seem to be invisible to diffraction-contrast examination, due to its short Burgers vector.

Because the $YBa_2Cu_2O_7$ cell is rhombic, there are 6 different types of orientational relationship, with the misfit varying for each one. At the first minimum (-45°), the energies for 2 different orientations of the epitaxial layer are almost equal. The growth of an epitaxial layer on a substrate having that orientation is expected to lead to polysynthetic twinning of that layer. For the minima at 27 and 61°, the misfits of the predominant orientational relationships are less than 0.1%, while the energies of competing orientations are larger. This can lead to the growth of highly perfect twin-free epitaxial layers at those substrate orientations.

Dislocation configurations and residual strain-levels for two types of interface in heterostructural $YBa_2Cu_3O_7/BaTbO_3$ and $PrBa_2Cu_3O_7/BaTbO_3$ thin films could be explained[209] in terms of 0-lattice theory and the bonding of ions at the interface. Depending upon the deposition temperature of $BaTbO_3$ layers, 2 types of interface were found. At a deposition-temperature of 780C, the $YBa_2Cu_3O_7/BaTbO_3$ interface was type-I, where a Cu-O chain plane of $YBa_2Cu_3O_7$ or $PrBa_2Cu_3O_7$ faced a TbO_2 plane of $BaTbO_3$. The interface dislocations had a line direction of [110] or [110], and residual misfit could be detected within a layer close to the interface. For a deposition temperature of 500C, type-I and type-II interfaces co-existed. In type-II, a Cu-O plane of $YBa_2Cu_3O_7$ or $PrBa_2Cu_3O_7$ faced a BaO plane of $BaTbO_3$. The misfit dislocations here lay along the [100] and [010] directions, and the lattice was strongly distorted near to the interface.

The effect of lattice-matching between $YBa_2Cu_3O_{7-\delta}$ films and buffer layers and between buffer layers and MgO substrates was considered[210], where $SrSnO_3$ and $BaSnO_3$ were the buffers. The $SrSnO_3$ had a lattice constant which was closer to that of $YBa_2Cu_3O_{7-\delta}$ than was $BaSnO_3$. The $YBa_2Cu_3O_{7-\delta}$ films were grown onto these materials, at the optimum growth temperature, by means of pulsed laser deposition. The full-width at half-maximum φ-scan of the (102) plane of $YBa_2Cu_3O_{7-\delta}$ films on $SrSnO_3$ buffer layers was smaller than that of $YBa_2Cu_3O_{7-\delta}$ films on $BaSnO_3$ buffers. Those for 005 rocking-curves of $YBa_2Cu_3O_{7-\delta}$ films on $BaSnO_3$ buffer layers were smaller than that of $YBa_2Cu_3O_{7-\delta}$ films on $SrSnO_3$ buffer layers. The difference in surface resistance, between $YBa_2Cu_3O_{7-\delta}$ film on a $BaSnO_3$ buffer layer and $YBa_2Cu_3O_{7-\delta}$ film on a $SrSnO_3$ buffer layer, was very small. This suggested that the superconductivity was not controlled mainly by lattice mismatch in this region.

The in-plane lattice constants of a-axis oriented $YBa_2Cu_3O_x$ films which had been grown epitaxially onto $SrTiO_3$ were studied[211] by using an ion channelling technique. The distance between two barium atoms, in films which were up to 3500Å thick, was almost the same as the lattice constant of $SrTiO_3$. This meant that the films suffered internal strain. Relaxed regions were detected using <301> axial channelling measurements. The back-scattering yields for the <110> channelling and <301> channelling measurements indicated that strain-relief increased with increasing film thickness.

When thin films of YBCO superconductor were deposited[212] by *in situ* laser ablation onto a monocrystalline (001) MgO substrate at 750C, the orthorhombic $YBa_2Cu_3O_{7-x}$ phase and the tetragonal $Y_2Ba_4Cu_8O_{16}$ phase were found within the film thickness. The thin films grew with the c-axis perpendicular to the interface and with the a and b axes of the YBCO locked into 2 main orientations with respect to the MgO axes, being: [100]YBCO∥[100]MgO and [110]YBCO∥[100]MgO.

When epitaxially oriented thin films of $YBa_2Cu_3O_{7-x}$ were grown[213] onto the (100) face of MgO by metalorganic decomposition, the as-prepared films consisted of monocrystalline platelets which lay flat on the MgO surface. Most of the crystallites were in perfect alignment of their c-axis with the [100]-axis of MgO, although some crystallites were misoriented by up to 7.5°. There was good epitaxial growth to within one lattice spacing of the MgO substrate, with a 0.51° spread in crystallite orientation.

When $GdBa_2Cu_3O_{7-x}$ thin films with a superconducting transition temperature of 89.6 to 92.18K, are grown[214] *in situ* onto a (001) $Zr(Y)O_2$ substrate using direct-current magnetron sputtering, the relationship between the a- and b-axes of the $GdBa_2Cu_3O_{7-x}$ thin films and the in-plane lattice vectors of $Zr(Y)O_2$ are such that there are 2 types of twin: 45° and 90°.

The 0-lattice theory was used[215] to analyse the structures of symmetrical tilt grain boundaries having a {001} rotation axis and to provide a theoretical interpretation of experimentally observed structures, of a near-Σ5 grain boundary in MgO, in terms of structural units and a periodicity of the 0-points at the boundary. Generalised decomposition formulae were derived for symmetrical tilt grain boundaries, and these were closely related to the distribution of irreducible rational numbers; again indicating a profound connection to the geometry-of-numbers.

When yttrium-aluminium perovskite/yttrium-aluminium monoclinic and yttrium-aluminium monoclinic/yttria eutectics were studied[216] using transmission electron and optical microscopy, 6 orientational relationships were identified in the former case and 2 in the latter. Near-coincidence site lattices were deduced for each eutectic, and orientational relationships were predicted on the basis of energy minimization. Predictions were also made using symmetry overlap criteria and an edge-to-edge model. All of the common orientational relationships had coincident low-index real-space lattice vectors which were consistent with the predictions of the edge-to-edge model. Three of the relationships involved matching planes which were co-zonal with coincident real-space directions that were consistent with the edge-to-edge model.

When $LuNi_2B_2C$ thin films were deposited[217] by pulsed laser ablation onto various MgO crystal planes, all of the samples exhibited a good c-axis orientation, but rocking-curves around the $LuNi_2B_2C$ (004) reflection indicated large differences in the width and shape of deposits on differing substrates. It was noted that growth of the film was effectively governed by the substrate, even in the case of large lattice mismatches: almost perfect epitaxial growth was observed for the (110) MgO surface, while multiple in-plane orientations were observed on the (100) and (111) planes. All of the observed orientations could however be explained by a simple near-coincidence site lattice model.

When $LaNiO_3$ thin films were grown on monocrystalline (001) MgO substrates, 2 growth stages were observed[218]: a (001)-oriented $LaNiO_3$ layer grew initially with a uniform in-plane alignment. Then a (110) orientation, aligned in-plane with the (001) layer, appeared. A consistent explanation of the preferred in-plane orientations was provided by near-coincidence site theory.

Transmission electron microscopic studies of interphase boundaries between SiC and β-Si_3N_4 in composites showed[219] that the boundaries between small intragranular nitride precipitates and the surrounding carbide grains were relatively free from intergranular films. On the other hand, the boundaries between large nitride grains and adjacent carbide grains were always coated with a thin intergranular film. An orientational relationship of the form, $[110]SiC\|[00\bullet1]Si_3N_4$; $(001)SiC\|(10\bullet0)Si_3N_4$, predominated between 3C SiC grains and intragranular β-Si_3N_4 precipitates, but there was no sign of a favoured orientational relationship existing between large nitride grains and adjacent carbide grains. The reason for the existence of an orientational relationship in the absence of an intergranular film was explained in terms of near-coincidence site lattice theory, leading to the discovery that the predominant orientational relationships between carbide grains and intragranular nitride involved the smallest misfits among all of the other possible orientational relationships.

A study[220] was made of photo-catalytic hydrogen release from water using a composite SiC/CdS catalyst having 2 types of hetero-interface. Type 1 consisted of hexagonal SiC and hexagonal CdS. Type 2 consisted of hexagonal SiC and cubic CdS. The type-1 composite led to an evolution rate which was 4 times higher than that of type 2, in spite of the similar band-gaps and electropotentials of cubic and hexagonal CdS. Lattice-matching was thought to occur between the 2 hexagonal compounds when the lattice-constant relationship was $3a_{H-CdS} = 4a_{H-SiC}$. The hexagonal SiC surface (006) also had an affinity for the hexagonal CdS (002) facet, due to their polar properties. These results demonstrated that lattice-matching plays an important role in forming efficient heterojunctions.

Density functional theory, with 3 different functionals, was used[221] to compute relevant terms for TiO_2 anatase and rutile film growth on the low-index surfaces of $SrTiO_3$ and $BaTiO_3$ cubic perovskites. The volumetric strain and areal substrate/film interface energies were calculated for (001)anatase‖(001)BaTiO$_3$, (001)anatase‖(001)SrTiO$_3$, (102)anatase‖(011) BaTiO$_3$, (102)anatase‖(011)SrTiO$_3$, (100)rutile‖(111)BaTiO$_3$, (100)rutile‖(111)SrTiO$_3$, (112)anatase‖(111)BaTiO$_3$ and (112)anatase‖(111)SrTiO$_3$ coherent interfaces. The terms were incorporated into a standard model for epitaxial nucleation, resulting in a reasonable agreement between experiment and density functional theory predictions of the preferred epitaxial polymorph. Those predictions of epitaxial stability were effectively independent of the 3 functionals used, and the results confirmed a proposed 20kJ/mol stability window for the prediction of those polymorphs which can be epitaxially stabilized.

When the grain-boundary structure of a TiO_2 rutile bi-crystal with a [001] symmetrical tilt boundary having a tilt angle of 66° and a consequent 1.4° deviation from an exact Σ13a to Σ17a relationship was studied[222], the deviation could not be accommodated by displacement shift complete dislocations. Other structural elements instead had to be introduced. The boundary was almost straight[223], with no step structures, but part of the boundary contained facet structures which consisted of low-index planes such as {310} and {110}. Contrast features due to high strain fields nevertheless existed on the grain-boundary plane, with a weak-beam dark-field spacing of 7.6nm. The strain was attributed to a distorted Σ13 unit structure, which was similar in structure to Σ17. The boundary consisted of a periodic array of mainly Σ13, and partially Σ17-like, unit structures. That is, the misfit angle was accommodated not by introducing secondary dislocations, but by a transformation of the basal unit structure.

On the basis of a lattice-matching procedure, optimum substrate combinations were assessed[224] for the coherent epitaxy of metal-organic frameworks. A detailed investigation of the growth of (011) metal-organic framework-5 on (110)TiO_2 (rutile) showed that the lowest-energy interface configuration involved a bi-dentate connection between two TiO_6 polyhedra, plus the de-protonation of terephthalic acid to a bridging oxide site.

Another computational scheme for selecting optimum substrates for epitaxial growth has been based[225] upon first-principles calculations of formation energies and elastic strain energy, together with topological information. As an example, a study was made of the stabilization of metastable VO_2 compounds. Lattice-matching and energy considerations led to the strategy of using homostructural growth on TiO_2 substrates. The VO_2 brookite phase would then be preferentially grown on the a-c TiO_2 brookite plane while columbite and anatase structures would favour the a-b plane of the respective TiO_2 phases. In general, a model which involved geometrical unit-cell area-matching between the substrate and the film, together with the resultant strain-energy density of the film, yielded qualitative agreement with experiment with regard to the heterostructural growth of known VO_2 polymorphs.

Given that hetero-epitaxial nanocrystals constitute an important class of materials, a reliable predictive method is required in order to predict interface planes while allowing for the existence of multiple orientational relationships and high-index faceting planes. One powerful technique is based[226] upon an invariant deformation element model and the fundamental crystallography of diffusional phase transformations. It has been applied to a complex metal-oxide nanocrystal system which exhibits up to 5 orientational relationships for 3 different growth orientations.

Dimer nanocrystals were synthesized[227] by growing In_2O_3 epitaxial material onto seed nanocrystals of UO_2 or FePt, and lattice strains were found to exist in both of the dimer nanocrystals. The In_2O_3 lattice expanded in UO_2/In_2O_3 dimers whereas, in the case of FePt/In_2O_3 dimers, the In_2O_3 lattice was compressed. The crystallographic orientation of the attachment of these dimers was identified, and a special Miller index was introduced in order to describe the crystallographic orientation of these hetero-dimer nanocrystals. An empirical law was proposed for the determination of the crystallographic attachment orientation in hetero-dimers. Rather than growth on the facets of seed nanocrystals, where lattice-mismatch is minimized, it was suggested that growth of an epitaxial material instead chooses those crystal facets for which the initial atomic monolayer of the material has the strongest affinity.

Cubic $Fe_{0.64}Ni_{0.36}$ precipitates were found[228], in a cubic $(Ni,Zn)Fe_2O_4$ matrix, which were in the form of an almost-periodic array of nanowires. The latter grew endotaxially, with corresponding cubic axes of the two structures almost perfectly parallel to one another; the deviation being smaller than $1°$. The interfaces between the matrix and the precipitates were well-defined crystallographic planes that were mainly arranged parallel to $\{111\}$. The nanowires were considered to be a self-organized nanostructure.

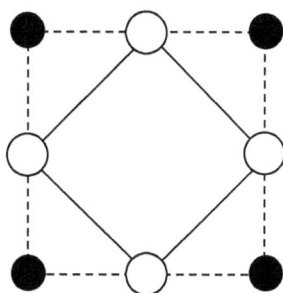

Figure 27. Proposed epitaxial matching for PbS (black) on rutile (white)

Aligned lead sulfide nanowires having a hyper-branched morphology have been produced[229] epitaxially on monocrystalline NaCl, (001) rutile and muscovite mica by chemical vapour deposition. An epitaxial match to the (100) plane of PbS was found for all of the substrates and, in addition, an epitaxial match to the (111) plane of PbS was also found in the case of mica. Sulfide nanowires of pine-tree form could be created only non-epitaxially in the presence of epitaxial hyper-branched clusters. This difficulty suggested that epitaxial growth might not be suitable for the creation of the dislocations which drove the formation of pine-tree nanowires. The mechanism of PbS epitaxial growth on NaCl was essentially self-explanatory, given that they have identical rock-salt structures and similar lattice constants. The upright orientation of orthogonally-branched nanowire clusters suggested that epitaxy on rutile involved the (100) planes of both materials. The fact that the same orientation of nanowire clusters was observed on rutile (001) surfaces indicated that the substrate surface was epitaxial with the (100) plane of PbS. Rutile (TiO_2) has a tetragonal structure and a unit-cell with a square (001) face, and dimension of 4.59Å, which constitutes the surface of the substrate used. The diagonal of the unit-cell face matches the lattice parameter of PbS to within 9%. A unit cell of rutile can therefore just fit inside the (100) face of a unit cell of PbS (figure 27). The epitaxial match of mica with both the (100) and (111) faces of PbS is intriguing. This is especially so in the case of (100) because the surfaces have different symmetries. The mica structure consists of alternating layers of positively-charged potassium ions and negatively-charged aluminosilicates which comprise a layer of face-sharing AlO_6 octahedra, sandwiched together with 2 layers of SiO_6 octahedra. Cleavage can be expected to occur where the bonding is weakest: between the potassium and aluminosilicate layers.

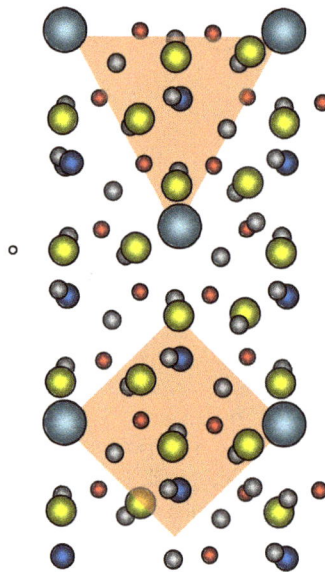

Figure 28. Proposed epitaxial matchings (shaded areas) for PbS on cleaved mica. Light blue: surface potassium atoms, yellow: deeper-lying aluminium atoms, grey: deeper-lying oxygen atoms, dark-blue: deeper-lying potassium atoms, red: deeper-lying silicon atoms

The structure of the potassium and silicate networks is essentially hexagonal. Because the diagonal of the (100) face of the PbS unit cell is within 7% of the distance between second-nearest neighbour potassium ions, it is possible for PbS to interact with them to give the epitaxial matches indicated by the triangles and squares in figure 28, for PbS (111) and PbS (100). The absence of a third form of epitaxy on cleaved mica has been attributed to the possibility that interaction between PbS and the deeper-lying potassium ions may prevent it.

Metalorganic chemical vapor deposition permits[230] the epitaxial growth of ZnTe thin films on substrates such as monocrystalline graphene and freshly-cleaved (001) mica in spite of the large in-plane lattice mismatches of some 75 and 17%, respectively. X-ray pole-figure analysis reveals preferred epitaxial alignments, with the out-of-plane orientation lying along [111] for the ZnTe film in both cases. For ZnTe on graphene, there is a primary in-plane orientation of [$\bar{1}$10]ZnTe||[1$\bar{1}$]graphene, plus 2 secondary in-plane orientations which are rotated by ±25.28° away from the primary domain. A geometrical superlattice areal mismatch model could be used to explain the existence of these primary and secondary domains for ZnTe on graphene. In the case of ZnTe on (001) mica, a single in-plane orientational domain was found: [$\bar{1}$10]ZnTe||[100]mica. In this case however, the predicted domain orientation between ZnTe and mica(001) which was based upon the geometrical superlattice areal mismatch model deviated slightly from the observed one. Multiple-order twinning domains of the primary domain were observed for both substrates, with the twinning domains remaining epitaxially aligned with respect to the substrate.

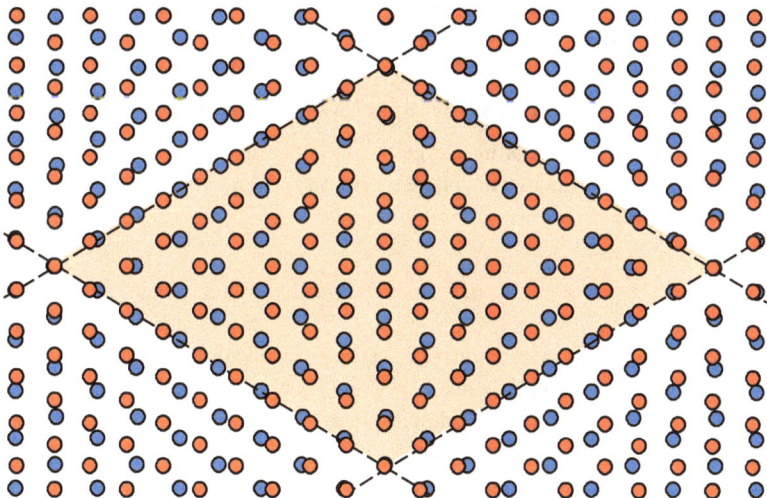

Figure 29. Common superlattice between InN (00•$\bar{1}$) and AlN (00•$\bar{1}$) planes

Again in spite of a lattice mismatch of 23% and a lack of high in-plane symmetry in the surface, epitaxial germanium films with a [111] out-of-plane orientation have been obtained[231] on mica. X-ray pole-figure analysis showed that multiple rotational domains existed in the epitaxial germanium film, with predominant in-plane orientations, between [$\bar{1}$10]Ge and [100]mica, of (20n)°; with n equal to 0, 1, 2, 3, 4 or 5. A superlattice areal mismatch model was again used to account for the probability of the in-plane orientation, and it was found to be qualitatively consistent with the observed predominant orientations.

High-quality heterostructures of pairs of materials can be formed[232] in the presence of so-called magic ratios between their lattice constants, even if those constants are very different[233]. Thus in the case of Si/Si_3N_4 the magic ratio is 2:1 and, in the case of AlN/Si, the ratio is 5:4. In the present case, of InN on AlN, the ratio is 8:9. Lattice-matching permits the formation of commensurate, and almost strain-free interfaces, having a common two-dimensional superlattice (figure 29). In the presence of nitrogen polarity, a pseudomorphic-to-commensurate lattice transition occurs within the first monolayer of growth and results in the appearance of a sharp heterojunction at the atomic scale.

Thin films of highly c-axis oriented aluminium nitride have been deposited[234] onto molybdenum by using AlN interlayers, giving layers of the overall form, AlN/Mo/AlN/Si. The crystallinity and orientation of the interlayers depended[235] upon their thickness, and markedly affected those of the molybdenum and AlN films. The latter films consisted of columnar grains and had a fiber texture. The local epitaxial relationship was:

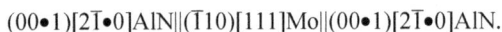

$$(00\bullet1)[2\bar{1}\bullet0]AlN\|(\bar{1}10)[111]Mo\|(00\bullet1)[2\bar{1}\bullet0]AlN.$$

It was concluded that the AlN interlayer was effective in decreasing the crystallization energy of the molybdenum due to coherent hetero-epitaxial nucleation. It was interesting that the local hetero-epitaxial relationship did not satisfy the criteria for hetero-epitaxial growth.

When copper films were deposited[236] onto epitaxial HfN/Si substrates by dc magnetron sputtering, the epitaxial growth of (111) and (110) copper on epitaxial (111) and (001) HfN films, respectively, was confirmed. These combinations of deposit and substrate could be explained in terms of lattice matching (figures 30 and 31). Both epitaxial HfN films had essentially flat surfaces, despite their differing orientational planes, but the surface-roughness of (110) copper film on (001) HfN was about half of that of (111) copper film on (111) HfN. The difference in surface morphology between (111) and (110) epitaxial copper was attributed to the various factors which controlled epitaxial growth at the Cu/HfN interface.

Materials Research Forum LLC
https://doi.org/10.21741/9781644900475

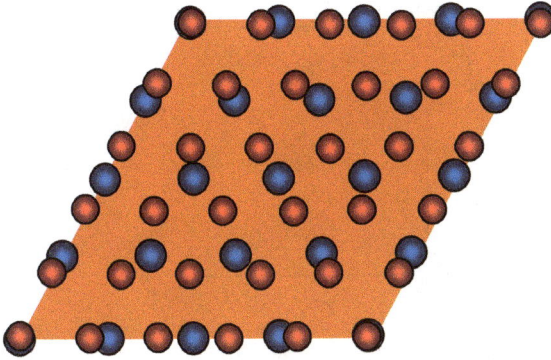

Figure 30. Superlattice cell consisting of 4 x 4 unit-cells
of (111)HfN (blue) and 5 x 5 unit-cells of (111) copper (red)

When (001)-oriented ruthenium films were prepared[237] on SiO_2/Si substrates by radio-frequency magnetron sputtering, using an O_2 flow-ratio which was lower than that required for RuO_2 formation, oxygen atoms were thought to act as a surfactant and to encourage the singly-oriented growth of ruthenium films. Highly (100)-oriented RuO_2 layers were subsequently deposited onto the ruthenium films via reactive sputtering. The (100)-oriented growth of a RuO_2 layer was proposed to originate from local epitaxial growth on (001)-oriented ruthenium grains because the mismatch between the superlattices of (001)Ru and (100)RuO_2 planes was small; the basic rationale being that: 3 x 4.499Å ~ 5 x 2.706Å (13.497Å ~ 13.530Å) in one direction and that 3 x 3.107Å ~ 4 x 2.343Å (9.321Å ~ 9.372Å) in the other direction (figure 32).

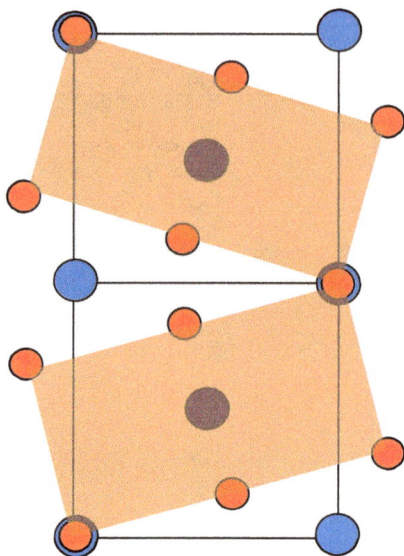

Figure 31. Diagonal-type epitaxy of a (110)Cu (red)
unit-cell on a (001)HfN (blue) unit-cell

X-ray diffraction diagnostics[238] of thin monocrystalline layers, on non-isomorphic substrates involving low-symmetry crystals and non-singular interfaces, permitted the determination of stress and strain tensors in the layer and an exact description of the orientational relationships. Study of the GaN/LiGaO$_2$ model heterostructure, which is associated with insignificant lattice mismatch, showed that the latter is equivalent to assuming a pseudomorphic nature for the epitaxial growth in the initial stages. The final state nevertheless corresponds to deep elastic-energy relaxation. The density of misfit dislocations in the system indicated that lattice mismatch was not the factor which controlled epitaxial-layer perfection.

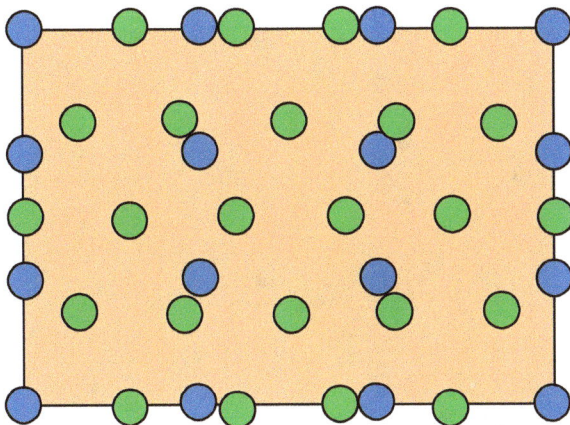

Figure 32. Epitaxial relationship between 5 x 4 (001)Ru and 3 x 3 (100)RuO₂ planes

When films of GaN which were about 0.4μm thick were grown[239] epitaxially onto a LaAlO₃ (100) substrate by using a molecular beam technique, the growth mode changed from island to lateral growth[240]. The orientational relationship between the GaN and the LaAlO₃ (100) substrate was [00•1]GaN‖[100]LaAlO₃; [01•0]GaN‖[0$\bar{1}$1]LaAlO₃, and [2$\bar{1}$•0]GaN ‖[011]LaAlO₃. The mismatch (figure 33) was less than 3% for the [0$\bar{1}$1] plane, and was thus much less than the 14% mismatch of the GaN/Al₂O₃ system.

Thin films of the form, (001)SrRuO₃, (001) CaRuO₃ and (205) BaRuO₃, were grown[241] epitaxially onto (001) LaAlO₃ substrates by means of laser ablation, with the aim of examining the effect of lattice-matching upon electrical conductivity. The (001) SrRuO₃ and (001) CaRuO₃ thin films exhibited terraces, with an orthogonal step structure, while (205) BaRuO₃ films exhibited an orthogonal structure, with tetragonal grains. The epitaxial thin films displayed metallic conduction, with (001) CaRuO₃ films having the highest (1.5 x 10⁵S/m) conductivity among the (001) SrRuO₃, (001) CaRuO₃ and (205) BaRuO₃ films. The higher conductivity was associated with a lower misfit between the thin film and the substrate.

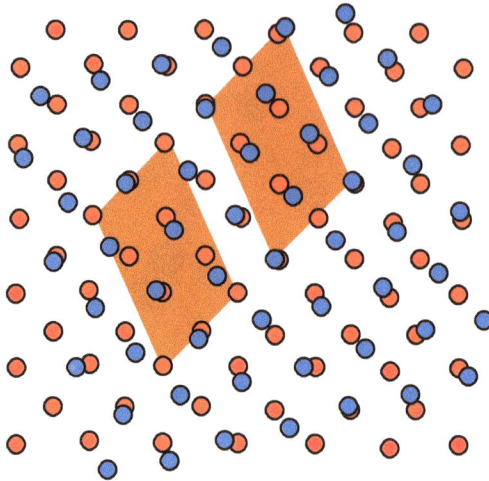

Figure 33. Model of lattice-matching between (00•1)GaN (blue) and (100)LaAlO₃ (red)

Nickel films which were 200nm-thick were grown[242], using dc magnetron sputtering, onto MgO (100) substrates at 20 to 700C. At 20C, a complex texture with predominantly <022>-orientated grains, co-existing with <1$\bar{4}$1> and traces of <002>, resulted[243]. At 100 to 200C, smooth monocrystalline layers with a completely <200> texture were obtained. Higher deposition temperatures always produced a <75$\bar{1}$> texture which was fourfold degenerate and twinned. This gradually became more defined at higher temperatures. There was also a transition to a faceted surface. An increase in deposition-temperature permitted already-nucleated nickel islands to rearrange into the <002> texture. At even higher temperatures, the nickel-atom mobility was higher, permitting the nickel to rearrange further and grow with a <75$\bar{1}$> texture. Such a texture accommodated strong nickel bonds and placed each interfacial nickel atom in a position which ensured minimal mismatch between the two lattices (figure 34). The (75$\bar{1}$)-plane nickel atoms formed in effect a c(3 x 1) surface lattice on the MgO surface. The surface-feature size and roughness increased by an order of magnitude as the crystallographic orientation changed from single-<002> to multiple-<75$\bar{1}$>. In other studies, the nickel films had a strong <111>-oriented texture, with high out-of-plane and in-plane orientations. At 300C the Ni$_{111}$ rocking-curve width was about 0.25°, indicating high crystalline quality. The films were built up from 2 domains for temperatures of up to 200C and at 600C while, at 300 and 400C, only a single domain formed.

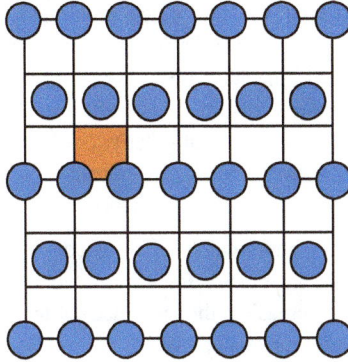

Figure 34. Lattice-matching of a MgO substrate and a (75$\bar{1}$) nickel layer
The orange square indicates one MgO unit cell

In a parallel study, epitaxial nickel films were sputtered[244] onto MgO (001) surfaces at 100 or 400C. The films which were deposited at the lower temperature were mainly [001]Ni‖[001]MgO and (010)Ni‖(010)MgO in orientation and were smooth. The higher-temperature deposited films were mainly of [75$\bar{1}$]Ni‖[001]MgO and ($\bar{1}1\bar{2}$)Ni‖(100)MgO in orientation and were facetted. The [75$\bar{1}$] orientation was four-fold degenerate and, for each of those 4 orientations, there was also a twin orientation; reflected about the (100)MgO planes and giving 8 possible orientations for nickel crystallites.

Thin films of LaAlO$_3$ and NdAlO$_3$ were deposited[245] onto biaxially-textured (100) nickel substrates by using a solution-deposition technique. Upon heating to 1150C in Ar-4%H$_2$, epitaxial films were formed which exhibited an out-of-plane as well as in-plane alignment For both films there were 2 in-plane orientations, the main one being (001)[100] and the minor one (001)[110]. The grain alignment in the films was approximately that of the alignment of the nickel substrate.

Studies of chemical vapour deposited Si$_3$N$_4$ on the (111) surface of SiC monocrystalline substrates revealed[246] 2 imperfect orientational relationships of the form: (10•0)Si$_3$N$_4$‖(111)SiC. They were imperfect because rotations of 2 to 6° commonly existed between the directions and planes involved. Possible atomic models for the interface exploited similarities between the [SiN_4] and [SiC_4] tetrahedra of the corresponding material. That is, the observed orientational relationships appeared to be due to a matching of the tetrahedra and lattices across the interface.

Arguments based upon electronic structure calculations and ionic radii considerations suggest[247] that interface charges make a contribution to the ionic part of lattice mismatch. This implies that the silicon-terminated 6H-SiC(00•1) surface is a better substrate for GaN than is the carbon-terminated surface. This combines the idea (Sasaki-Matsuoka), that substrate-quality depends upon electrical polarity, with the usual assumption that the best substrate is one which exhibits zero lattice-mismatch with respect to a deposited film.

Heterojunction $SrTiO_3\|NaTaO_3$ photo-catalysts were constructed[248] by hydrothermally decorating $NaTaO_3$ nanocubes with $SrTiO_3$ nanoparticles. The structures of these perovskites are similar, thus increasing the interface lattice-match and promoting the migration of photo-generated carriers between the $SrTiO_3\|NaTaO_3$ interfaces. In comparison to pristine $NaTaO_3$ and $SrTiO_3$, the composites had a markedly increased ability to degrade rhodamine-B under ultra-violet light. The partial replacement of O^{2-} by N^{3-} in the TaO_6 octahedron further narrowed the band-gap of $NaTaO_3$ and thus markedly increased the photo-catalytic ability of the $SrTiO_3/NaTaO_3{:}N$ heterojunction under visible light.

In an interesting inversion of the usual priorities, coherent $In_xAl_{1-x}N$ (x = 0.15 to 0.28) films were grown[249], by metalorganic chemical vapor deposition, onto GaN templates in order to investigate the effect of the resultant stresses upon wafer curvature. In the case of $In_{0.18}Al_{0.82}N$, there was no change in the curvature; thus confirming that films of this composition were latticed-matched to GaN, in accord with Vegard's law. The tests thus served in effect to confirm obedience to Vegard's law rather than using composition-variation to control film stability.

The exchange-bias properties of 5nm Co/CoO ferromagnetic/antiferromagnetic core/shell nanoparticles which were highly dispersed in a Cu_xO matrix were optimized[250] by matching the lattice parameter of the matrix to that of the CoO, leaving a 0.3% lattice-mismatch with respect to the shell. This small mismatch structurally stabilized the CoO and favoured the existence of a large fraction of uncompensated moments in the shell; which then increased exchange-bias effects. It was concluded that lattice-matching could be a very means for improving the exchange bias properties of core/shell nanoparticles.

An epitaxial relationship, with long-range lattice-matching, could be established between the (100) plane of the bismuth-layered structure, $Bi_4Ti_3O_{12}$, and the (00•1) plane of wurtzite-structured ZnO[251]. That is, (100)[001]$Bi_4Ti_3O_{12}\|$(00•1)<01•0>ZnO. Epitaxial

(100)-oriented holmium-doped $Bi_4Ti_3O_{12}$ thin films having the composition, $Bi_{3.6}Ho_{0.4}Ti_3O_{12}$, were integrated with (00•1)-oriented aluminium-doped ZnO layer buffered c-sapphire substrates using pulsed-laser deposition, with the epitaxial relationship: (100)[001] $Bi_{3.6}Ho_{0.4}Ti_3O_{12}\|(00•1)<01•0>ZnO$. The hetero-epitaxy of (100) $Bi_4Ti_3O_{12}$ on (00•1) ZnO, with long-range lattice-matching, offered the possibility of creating $Bi_4Ti_3O_{12}$/ZnO ferroelectric wide-bandgap semiconductor heterostructures.

A density functional theory study[252] of graphene/hexagonal-BN thin films showed that the interlayer interaction-energy of a monolayer film was inversely proportional to the layer-number. The analysis was based upon a simulation of the interlayer interactions in lattice-mismatched thin films and showed that films having 4 or more layers could possess stable lattice-matched stacking geometries. The maximum value of the band-gap of lattice-matched films having a given layer number, but differing stacking sequences, decreased as a function of the layer number. The band-gap could also be tuned by applying an external electric field. It was possible to envisage the creation of a 6-layer graphene/h-BN bilayer thin film with a 99meV band-gap, or a graphene-like linear dispersion, depending upon the stacking sequence chosen.

The most stable matching of silicon carbide and graphene has been found[253] by using molecular dynamics calculations, and genetic algorithms have been shown to offer a more efficient approach than do conventional structural optimization methods when searching for the best matching of lattices.

Due to the large lattice mismatch of some 20% which theoretically exists between SiC and silicon, it might be thought difficult to grow usable monocrystalline Si/SiC heterojunctions[254]. Silicon films with a <111> preferred orientation have been deposited[255] onto the 6H-SiC(00•$\bar{1}$) C-face. These films had an epitaxial relationship to the 6H-SiC substrate, and the parallel-plane alignment was: (111)Si$\|$(00•$\bar{1}$)6H-SiC. Misfit dislocations at the Si/6H-SiC interface accommodated most of the lattice-mismatch strain. Along the interface, every 4 silicon (111) lattice planes matched up with 5 6H-SiC (00•$\bar{1}$) lattice planes. On the basis of this 4:5 lattice-matching, the lattice structure of the Si/6H-SiC interface and its energetics were investigated using molecular dynamics simulations. When the silicon films grew preferentially along the <111> orientation on the 6H-SiC (00•$\bar{1}$)C-face, the misfit strain in the silicon layer was markedly reduced, due to a relaxation of the carbon atoms in the SiC layer close to the Si/6H-SiC interface. The Si/6H-SiC heterostructure thus had a stable interface, with an interface formation energy of -14.24eV. The 4:5 matching of atomic matching structures had already been observed in the case of the (111)Si$\|$(001)6H-SiC interface. It was noted then[256] that there was a minor lattice mismatch of 0.26%. Molecular dynamics simulations predicted that the interfaces were quite stable, with a formation energy of - 22.452eV.

Monocrystalline $KY_{(1-x-y-z)}Gd_xLu_yYb_z(WO_4)_2$ layers have been grown[257] onto undoped $KY(WO_4)_2$ substrates by means of liquid-phase epitaxy, with the co-doping of Gd^{3+} and Lu^{3+} being used to lattice-match the layers. A wide range of Yb^{3+} concentrations could be obtained by replacing the Lu^{3+} ions. Up to a maximum lattice mismatch of about 0.08%, crack-free layers could be grown by systematically varying the Y^{3+}, Gd^{3+}, Lu^{3+} and Yb^{3+} concentrations. The lattice constants and other properties of the co-doped layers could be predicted from weighted averages of the values of those properties for the pure compounds.

Optical waveguide films can be prepared[258] by growing barium sodium niobate onto potassium titanyl phosphate substrates using pulsed laser deposition, with the (110) plane of barium sodium niobate being parallel to the (001) plane of the phosphate and with a small lattice mismatch of 2.9%. An effective refractive index and effective thickness of the niobate films, caused by Goos-Hanchen shifts, is found.

Lattice-Matching to Metals

Local minima were found[259] in the dependence of the energy of interphase boundaries upon the dimensional and orientational misfit of special interphase boundaries in metal systems with a face-centerd cubic lattice and (111)/(001) or (111)/(110) boundaries. Criteria were found which permitted the prediction of orientational relationships that corresponded to these special boundaries. The interphase boundaries were found to exhibit a periodic sub-structure over the entire range of dimensional and orientational misfits. The Burgers vectors of the interfacial dislocations could be both lattice vectors, and vectors of the displacement shift complete lattice for the corresponding coincidence site lattice.

Epitaxial growth of cerium on vanadium was achieved[260] by using molecular beam epitaxy and was characterized using high-energy electron and X-ray diffraction. This showed that the cerium phase was contracted by about 8% in the basal plane and was expanded out of the plane by 2%, as compared with ambient-temperature and ambient-pressure face-centerd cubic cerium. The relative orientation of the (111) cerium and (110) vanadium planes was different to the experimentally and theoretically expected result. This suggested that electronic factors, going beyond simple geometrical lattice-matching considerations, were important. For example, metallic superlattices exhibit a correlation between the epitaxial growth temperature and temperatures deduced from the equilibrium phase diagram[261]. This correlation, together with molecular dynamics simulations of epitaxial growth, indicates that thermodynamics plays an important role in the growth of epitaxial films; especially with regard to the sharpness of the interface.

The energy of a face-centerd cubic (111) overlayer on a body-centerd cubic (110) substrate was calculated[262] by using empirical Morse potential functions to represent the two-body interactions. In the cases of a nickel or copper overlayer on a tungsten substrate, the Nishiyama-Wassermann orientation, $(0\overline{1}1)fcc\|(001)bcc$, was the most stable one. A gold or silver overlayer on a tungsten substrate found its energy minimum in the Kurdjumov-Sachs orientation: $(110)fcc\|(1\overline{1}\overline{1})bcc$. A lead overlayer on tungsten was not associated with any clear energy minimum. All of these results were in good agreement with moiré pattern studies and with the predictions of 0-lattice theory. The existence of a vertical displacement between the body-centerd cubic and face-centerd cubic lattices did not change the epitaxial orientation.

The 0-lattice theory has been used[263] to clarify the connection between interphase boundary fracture and deviations from the Kurdjumov-Sachs relationship for $Ni_3Al(\gamma')$ precipitates at the grain boundaries of $NiAl(\beta)$. In β-phase bi-crystals having controlled orientations, one variant of γ'-film was chosen so as to satisfy the Kurdjumov-Sachs relationship with respect to a neighbouring β-phase crystal, but to deviate from the relationship with respect to an adjacent β-phase crystal. During tensile deformation at room temperature, fracture occurred preferentially at interphase boundaries which deviated from the Kurdjumov-Sachs relationship. There was no plastic deformation, and the fracture stress decreased with increasing deviation-angle. When the Kurdjumov-Sachs relationship was present, there was slip transfer, from the γ'-film to the β-phase crystal, across the coherent interface.

A combination of 0-lattice, good matching site, constrained coincidence site lattice and constrained complete pattern shift lattice models has, by the way, been used[264] to analyze a newly discovered orientational relationship between Widmanstätten cementite precipitates and the austenite matrix of a 1.3C-14Mn steel.

Atomistic approaches were used[265] to model the structures of various named (Bagaryatskii, Isaichev, Pitsch-Petch) orientational relationships between the ferrite and cementite components of pearlite. Dislocation arrays were associated with all of the relationships, and their spacing and direction depended upon the lattice mismatch. For a given orientational relationship, the interfacial chemistry affected only the interfacial energy and not the dislocation behaviour. The presence of similarly sited pairs of iron atoms, and the closeness of carbon atoms to the interface also affected the interfacial energy. The Isaichev relationship was the most energetically favorable one, together with nearby deviations from it. The interfacial-energy approximations which resulted from using the atomistic approaches could be reproduced by using an anisotropic continuum model which was based upon 0-lattice theory. The latter could also characterize the Burgers vectors at the interface plane which lay in high-symmetry directions of the ferrite.

This O-lattice theory plus anisotropic continuum theory had already explained the energetics and structure of the Bagaryatskii relationship between the ferrite and cementite components of pearlite. The atomistic approach had shown[266] meanwhile that the interface consisted of a rectangular dislocation-array, aligned with high-symmetry directions in the interface. The interface could be constructed by using 3 atomic planes, in the cementite structure, which governed the chemistry and registry of the interface and thus controlled the interfacial energy. The FeC/Fe terminating plane was the lowest-energy one because interfacial dislocations could extend most easily on those planes and thereby reduce the interfacial energy.

The prediction of edge-facet details was examined[267], with regard to α-precipitates in the β-phase matrix of a Ti-7.26wt%Cr alloy, by using a near-coincidence-site model combined with an analysis of moiré planes based upon 0-lattice theory. The distribution of near-coincidence-site clusters exhibited 2 orders of modulated periodicity, which were defined by the approximate intersections between several sets of moiré planes. Two possible edge-facets could be suggested, based upon the near-coincidence-site distribution. The calculated orientations and dislocation structures of edge-facets, based upon misfit analysis, were consistent with experimental observations.

Two intermetallic phases were found[268] in composite quaternary alloys containing 21 to 22vol% of Al_3M, where M was titanium, vanadium or zirconium. One phase was vanadium-rich $D0_{22}$ $Al_3(Ti,V,Zr)$ and the other was zirconium-rich $D0_{23}$ $Al_3(Ti,V,Zr)$. The lattice constants of the precipitates depended sensitively upon the transition-element contents and tended to obey Vegard's rule. The lattice misfit between an Al_3M phase and the matrix did not change greatly, while the lattice misfit between $D0_{22}$ and $D0_{23}$ Al_3M phases decreased, with increasing titanium content, to such an extent that the interface between the phases became difficult to distinguish.

High-resolution electron microscopy of γ-$Mg_{17}Al_{12}$ particles in heat-treated (473K, 8h) AZ91 alloy revealed[269] particles having a Pitsch-Schrader orientational relationship to the matrix. Their habit plane and growth direction could be explained in terms of the 0-lattice, constrained coincidence site lattice and constrained complete pattern shift lattice models.

The 0-lattice theory was used[270] to explain the orientational relationship and morphology of plate-like $Al_6(Mn,Fe)$ dispersoids in AA5182 alloy. The orientation of the precipitate (p) with respect to the matrix (m) was: $(001)p\|(3\bar{1}5)m$; $[\bar{1}10]p\|[21\bar{1}]m$, and the habit planes were (001)p and ($3\overline{1}5$)m. Atom-atom matching occurred in good-matching regions between dislocations on the habit plane.

The non-magnetic alloy, CoRu, was studied[271] as a matching-layer for the epitaxial growth of a layer of magnetic $CoCr_{22}Pt_{14}B_4$, with its hexagonal close-packed (110) orientation on CrTi alloy underlayers with their body-centerd cubic (100) orientation. The CoRu matching-layer led to an in-plane c-axis orientation, (110), of the CoCrPtB layer which had the same hexagonal close-packed structure as the CoRu and increased the magnetic coercivity due to its magnetocrystalline anisotropy. The addition of a combination of a $CrTi_{10}Mo_{10}$ or $CrTi_{10}W_{10}$ underlayer and a $CoRu_{40}$ layer could markedly improve the (110) orientation of the CoCrPtB layer and increase the coercivity as compared with that of $CrTi_{20}$ underlayer with a $CoRu_{40}$ layer.

High-resolution microscopy of germanium needle or lath precipitates in Al-Ge alloy has shown[272] that the many orientational relationships of twinned <100> and <110> needles are related to just 3 basic lattice correspondences. The feature which is common to the <100> and <110> needles is the arrangement of 5 co-zonal twin segments whose relative degree of development determines the cross-sectional morphology. Growth mechanisms which were based upon multiple twinning could account for the formation of all of the observed precipitate morphologies and defect sub-structures.

When superconducting $YBa_2Cu_3O_{7-x}$ films were grown[273] by laser ablation onto (001), (110) or (111) monocrystalline silver surfaces, X-ray diffraction showed that the films were always aligned with the crystallographic axes of the substrate and that these orientations were consistent with the predictions of near-coincident site lattice theory.

Interconnect metallizations consist[274] of sputtered stacks which typically comprise thin titanium and titanium nitride underlayers that are covered by aluminium-alloy. The textures of such films are an important factor. In one study, an increase in the substrate temperature caused a transition in the titanium texture from <00•2> to <10•1>. When TiN was reactively sputtered onto the titanium, the <00•2>texture induced a <111> TiN texture while the <10•1> titanium texture was related to the formation of a <311>TiN texture. The presence of only a titanium underlayer here sharpened the <111> aluminium texture but did not induce a textural component other than <111>Al. On the other hand, textured TiN in a Ti/TiN underlayer-stack led to so-called textural inheritance, in which <111>TiN → <111>Al and <311>TiN → <311>Al. The orientation of a film in a stack depended upon the surface energy of the top layer and upon the interfacial energies between the layers. It was concluded that the atomic bonding in the underlayer markedly affected minimization of the interfacial energy and governed whether lattice-matching between the underlayer and the top layer occurred (table 12).

Table 12. Lattice-matching between underlayer and overlayer films in multilayer stacks

Underlayer/Overlayer	a(Å)	b(Å)	A(Å^2)	Δa(%)	Δb(%)	ΔA(%)	Result
Ti(00$\bullet$2)[11$\bullet$0]‖Al(111)[1$\bar{1}$0]	2.95	5.11	15.07	12.94	12.94	15.88	C
Ti(10$\bullet$1)[1$\bar{2}$$\bullet$0]‖Al(111)[1$\bar{1}$0]	2.95	10.67	31.49	12.94	17.08	110.02	U
Ti(00$\bullet$2)[11$\bullet$0]‖Al(311)[01$\bar{1}$]	2.95	10.22	30.15	12.94	17.07	110.01	N
Ti(10$\bullet$1)[1$\bar{2}$$\bullet$0]‖Al(311)[01$\bar{1}$]	2.95	10.67	31.49	12.94	111.03	113.98	N
Ti(10$\bullet$1)[1$\bar{2}$$\bullet$0]‖Al(111)[11$\bar{2}$]	19.75	10.67	157.45	-0.87	-7.30	-8.17	L
Ti(10$\bullet$1)[10$\bullet$0]‖Al(311)[1$\bar{1}$$\bar{2}$]	35.40	10.67	377.87	12.94	17.08	110.02	N
Ti(00$\bullet$2)[11$\bullet$0]‖TiN(111)[1$\bar{1}$0]	2.95	5.11	15.07	-1.67	-1.67	-3.35	C
Ti(00$\bullet$2)[11$\bullet$0]‖TiN(311)[01$\bar{1}$]	2.95	10.22	30.15	-1.67	12.65	10.94	N
Ti(10$\bullet$1)[1$\bar{2}$$\bullet$0]‖TiN(111)[1$\bar{1}$0]	2.95	10.67	31.49	-1.67	14.27	12.59	L
Ti(10$\bullet$1)[1$\bar{2}$$\bullet$0]‖TiN(311)[1$\bar{1}$0]	2.95	10.67	31.49	-1.67	16.81	15.14	C
Ti(10$\bullet$1)[1$\bar{2}$$\bullet$0]‖TiN(111)[11$\bar{2}$]	14.75	10.67	157.45	-5.66	-12.39	-18.05	L
Ti(10$\bullet$1)[10$\bullet$0]‖TiN(311)[1$\bar{1}$$\bar{2}$]	35.40	10.67	377.87	12.66	12.66	15.32	L
TiN(111)[110]‖Al(111)[1$\bar{1}$0]	3.00	5.20	15.58	14.53	14.53	19.07	C
TiN(111)[110]‖Al(311)[01$\bar{1}$]	3.00	10.39	31.16	14.53	18.60	113.13	N
TiN(111)[112]‖Al(311)[01$\bar{1}$]	10.39	32.99	342.8	14.53	10.29	14.82	N
TiN(311)[011]‖Al(311)[01$\bar{1}$]	3.00	9.95	29.84	14.53	14.53	19.07	C
TiN(311)[011]‖Al(111)[1$\bar{1}$0]	3.00	9.95	29.84	14.53	10.29	14.82	U

a,b: unit-cell dimensions, A: matching superlattice area, Δa, Δb, ΔA: misfits between underlayer and overlayer, C: consistent with observations, U: unexplored, N: not observed, L: unlikely

Changes in C_{60}(111) in-plane structures on (111) gold substrates were investigated[275] at 120 to 290C. Epitaxial orientational re-ordering was observed at about 150C, and was attributed to a surface structural instability/reconstruction of the uppermost layer of the (111) substrate. Twinned in-phase structures predominated at lower temperatures. The close-packed directions of those structures were parallel to the close-packed directions of the non-reconstructed (111) gold substrate. That is, they were oriented in the R0.0 direction. Lower concentrations of minority-phase twinned in-plane structures, oriented at ±R30.0 were also observed to co-exist with the in-phase structures. The ±R30.0 minority-phase structures were commensurate with the non-reconstructed (111) substrate. No in-phase structures were observed at temperatures higher than the above re-ordering temperature, but commensurate R30.0 structures formed. The major phase consisted of structures which were double-positioned in the plane, ±R13.0 to ±14.2, with respect to the close-packed directions of the twinned commensurate structures. These double-positioned structures required large coincident unit-cells in order to achieve lattice-matching with non-reconstructed (111) gold substrates. The commensurate, R30.0, in-phase, R0.0, structures both grew with the bulk lattice spacings. Assuming that the surface of the substrate had the bulk structure and that the adsorbed C_{60} overlayers kept their bulk lattice spacings, the C_{60} structures could be lattice-matched almost perfectly to the non-constructed substrate, in the R30.0 orientation. The in-plane structures were then commensurate.

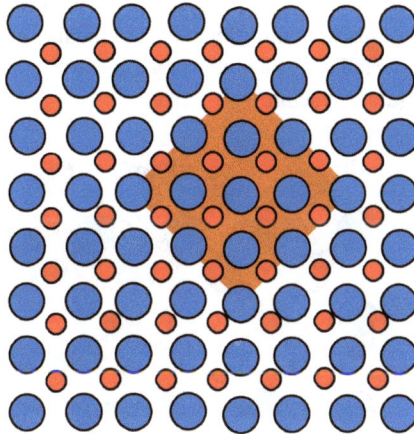

Figure 35. Extended ($2\sqrt{2}$ x $2\sqrt{2}$)R45° domain matching of [100]RE_2O_3 on [110]Ni

Assuming similar energies for the in-phase and competing double-positioning structures, those structures – unlike the commensurate one – could not be lattice-matched to the non-reconstructed substrate over small unit-cells. The lattice mismatches for the smallest unit-cells, across the interfaces of these structures, were about 13 and 4%. These mismatches were larger than those (circa 3.5%) normally found for strained-layer epitaxy. Because of this large disparity, the strained layers would have very different elastic energies. In order to lattice-match the in-phase structures, R0.0, to the non-reconstructed substrates, a 38 x 38 unit cell would be required. By using such a large unit cell, the C_{60} overlayer could be lattice-matched without suffering any strain. With increasing temperature, the in-plane orientations of those structures changed from ±R13.0 to ±R14.2. The in-phase, R0.0, and commensurate, R30.0, structures did not exhibit any dependence upon the substrate temperature or upon the structural quality of the substrate.

In a previous study of C_{60} growth on (111) silver substrates, the structures had been shown to consist of twinned commensurate forms growing with high-symmetry almost-perfectly lattice-matched orientations (2 3 x 2 3 ±R30° substrate unit cells) or forms in which close-packed directions were in lower-symmetry in-plane orientations. The latter forms could be lattice-matched to the substrates over larger unit-cells than could the commensurate forms. X-ray-diffraction data showed that their in-plane $(2\bar{2}0)$ mosaic peaks constituted double-positioning peaks with respect to the $[2\bar{2}0]$ directions of the twinned commensurate structures which lay between them. With increasing growth temperature, the close-packed directions of the double-positioning structures grew preferentially with orientations which converged towards the close-packed directions of the twinned commensurate structures.

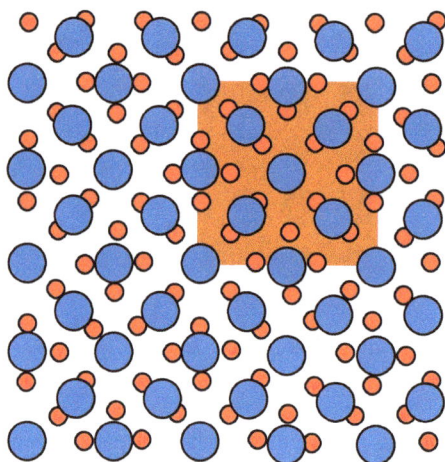

Figure 36. Extended (3 x 3) domain matching of [100]RE$_2$O$_3$ on [100]Ni

An investigation[276] of the structures and alignments of epitaxial Y_2O_3 and Gd_2O_3 films on cube-textured (001) nickel substrates was used to develop a general approach to the prediction of the morphologies of RE_2O_3 films on cube-textured nickel-based alloys (figures 35 and 36). The [100]-axis of RE_2O_3 oxides was found to prefer to align itself to the [110] nickel axis, while the [001]-axis was aligned with the [001] nickel axis. Such a $(2\sqrt{2} \times 2\sqrt{2})R45°$ extended domain matching appeared to be favoured due to a high coincidence-site density agreement with the (001) nickel surface atoms, in spite of large lattice-mismatches with respect to (001) nickel. On the other hand, a (3 x 3) extended domain matching was not favoured on (001) nickel in spite of relatively small lattice-mismatches with respect to (001) nickel. This was attributed to a low coincidence-site density. The matching preference varies from system to system (table 13).

Table 13. Lattice mismatches for various $(001)RE_2O_3\|(001)Ni$ systems and $(2\sqrt{2} \times 2\sqrt{2})R45°$ or (3 x 3) extended domains

RE_2O_3	a(Å)	(3 x 3) mismatch(%)	$(2\sqrt{2} \times 2\sqrt{2})R45°$ mismatch(%)
La_2O_3	11.380	7.765	14.303
Pr_2O_3	11.136	5.456	11.852
Nd_2O_3	11.048	4.621	10.968
Sm_2O_3	10.932	3.523	9.803
Eu_2O_3	10.866	2.898	9.140
Gd_2O_3	10.813	2.396	8.608
Tb_2O_3	10.728	1.591	7.754
Dy_2O_3	10.667	1.103	7.141
Ho_2O_3	10.607	0.445	6.539
Er_2O_3	10.547	-0.123	5.936
Tm_2O_3	10.488	-0.682	5.343
Yb_2O_3	10.439	-1.146	4.851
Y_2O_3	10.604	0.417	6.509
Lu_2O_3	10.391	-1.600	4.369
In_2O_3	10.118	-4.186	1.627
Cm_2O_3	11.000	4.167	10.486
Pu_2O_3	11.040	4.545	10.888
Sc_2O_3	9.845	-6.771	-1.115
Tl_2O_3	11.543	-0.161	5.896

Free-standing line-patterns of copper and nickel were produced[277] by applying photo-lithography and electrodeposition techniques to glass wafers covered in polycrystalline gold or amorphous Ni-P layers. Substrate-unaffected growth was observed[278] for copper lines on amorphous Ni-P, while highly-textured fine-grained gold layers strongly promoted the nucleation of copper crystallites having a preferred orientation. For certain pattern geometries, evidence was found for an epitaxial orientational relationship between copper and gold. Although conventional epitaxy failed for the copper-gold system, the experimentally observed orientational relationship could be viewed as involving a 30° rotation of (111)Cu around (111)Au (figure 37).

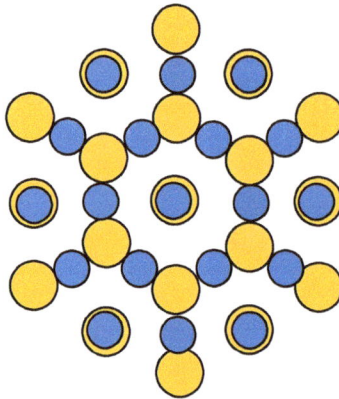

Figure 37. Possible lattice-matching between the (111) planes of copper (blue) and gold (orange), involving a 30°-rotation of the former lattice with respect to the latter

Monocrystalline graphene on an amorphous or crystalline SiO_2 substrate has been used[279] for studies of 100nm epitaxial copper films. X-ray diffraction and electron back-scattering diffraction showed that the copper films were always monocrystalline, with a (111) out-of-plane orientation and in-plane $\Sigma3$ twin domains of 60° rotation. The crystallinity of the SiO_2 substrates had a negligible effect upon the copper orientation during epitaxial growth, suggesting that the graphene exerted a considerable screening effect. When polycrystalline copper was instead grown epitaxially on polycrystalline monolayers of graphene having 2 orientation domains which were offset by 30° with respect to each other, the crystal orientation of the epitaxial copper film followed that of the graphene. The copper film thus consisted of 2 orientation domains which were also offset by 30° with respect to each other.

A proposed procedure for predicting precipitate facet-planes was based[280] upon finding those interphase boundaries, possessing a high density of near-coincidence sites, via a computer-search for lattice correspondences. In the case of face-centerd cubic to body-centerd cubic transformations, many lattice correspondences produced a similar degree of matching in three dimensions but very different degrees of matching in two dimensions. The (121) habit-plane of precipitates in Ni-Cr alloy, for example, was found to be a near-coincidence site boundary within which each near-coincidence site lattice-cell contained 3 atoms. Experimental observations suggested however that the selected habit-plane owed more to geometrical matching than to the existence of a lattice correspondence during growth.

The (334) Σ17b plane matching grain boundary of molybdenum is regarded as being one of low energy[281]. It exhibits a high fracture strength, in spite of its poor atom-matching when compared with that of the Σ9 and Σ11 coincidence boundaries. It is found that interface dislocations assure coherency at the boundary, in addition to displacement shift complete dislocations. The coherent nature of the boundary structure, due to these features, is considered to be responsible for the relatively high fracture strength of the boundary.

Model potentials of Heine-Abarenkov type were used[282] to calculate the energies of metallic bilayer systems such as Al/Li, Al/Mo and Al/Fe. The epitaxial relationships were determined by minimizing the energy of the various pairs of atomic layers. The results were in good agreement with the predictions of 0-lattice theory. The epitaxial relationships were of Nishiyama-Wassermann type, (112)fcc||(110)bcc, in the case of Al/Li and Al/Fe and of Kurdjumov-Sachs type, (110)fcc||(111)bcc), in the case of Al/Mo. Crystallographic features of zinc and zinc-nickel deposits on steel sheets were studied[283] using transmission electron microscopy and 0-lattice theory. The former results showed[284] that the hexagonal close-packed β-phase electrodeposits grew epitaxially on α-iron substrates, with a Burgers alignment which was in agreement with the predictions of 0-lattice theory, even allowing for the fact that the a-axis and c-axis dimensions varied as a function of the nickel content. The 0-lattice theory could also explain the cube/cube orientational relationship which was observed in the case of the nickel-rich β-phase.

Transmission electron microscopy was used[285] to determine the matrix/precipitate orientational relationship and similar features in Cr-Ni alloys. The use of 0-lattice theory showed that it was probable that the metastable 9R-phase was a transition-stage at lower reaction temperatures because lattice matching at the habit-plane between body-centerd cubic material and 9R was better than the matching at the habit-plane between body-centerd cubic and face-centerd cubic materials. The ability of a phenomenological theory of martensite crystallography to predict the habit-plane of 9R plates resulting from a diffusional process was attributed to the small lattice-invariant deformation which was required in order to produce an invariant plane in Cr-Ni alloy. Under these conditions, the habit-plane which was predicted by the martensite-related model almost coincided with the best-matching interface which was predicted by 0-lattice theory.

The martensite transformation has been analyzed[286] in terms of invariant-line and 0-lattice theory. The formation of martensite involves the migration of a well-defined (121) face-centerd cubic glissile interface, while its misfit dislocations produce a lattice invariant deformation. The latter is slightly retarded following the migration of the interphase, given that a thin plate-like zone can exist in the martensite near to the well-defined interface. When the temperature reaches the martensite-start point, the lattice parameter of the austenite matrix is $\sqrt{3}/2$ times that of the martensite. This critical condition for spontaneous transformation agrees with the alternative criterion that the stacking-fault energy in the matrix should be negative.

The phenomenological theory of martensite crystallography, and 0-lattice theory, were compared[287] with regard to the description of 0-line transformation strain. The displacement which was associated with the long-range strain component was shown to be identical in both treatments when the planes which were defined by the path of 0-lines were correlated with the dislocation slip plane. A feature of 0-lattice theory was that this displacement was permitted to vary, for a given habit plane. When the displacement was associated with an interfacial step, it was equivalent to a transformation dislocation.

By using 0-lattice theory, the orientational relationship, habit-plane orientation and matching or mismatching of certain sets of planes in the habit plane of the lath martensite of Fe-Ni-Mn alloy could be explained[288].

Molybdenum bi-crystals with symmetrical tilt boundaries, <001>(130) $\Sigma 5$, <001>(120) $\Sigma 5$, <001> (140) $\Sigma 17$, <001> (150) $\Sigma 13$, and <110> (332) $\Sigma 11$ coincidence boundaries were studied[289] using transmission electron microscopy. In pure specimens, periodic structures occurred along the boundary and could be well-explained by geometrical models such as coincidence site and 0-lattice theories. The grain boundary structures consisted of a combination of structural units involving stable boundaries. In impure specimens, precipitates and lattice bending were observed at the boundaries. It was concluded that impurities are one of the most important factors which affect boundary structures in molybdenum.

Grain-boundary interconnection has been defined as the matching of 2 crystallographic planes from adjoining grains, and to be the critical parameter which defines the nature and properties of a grain boundary[290]. Zone-melted 4N-purity aluminium was repeatedly forged multi-directionally to a true strain of 4 and recrystallized at 360C, thus producing

Materials Research Forum LLC

https://doi.org/10.21741/9781644900475

randomly oriented grains with an average size of about 30μm. The results showed that the grain-boundary interconnection for any group of grain boundaries having a given misorientation was not random but exhibited a marked preference for planes of low Miller-index which formed mixed and twist grain boundaries. Among high-angle boundaries, {111}‖{111} was the most frequent grain-boundary interconnection, this being due mainly to the grain boundaries formed by rotation around <111>, <122> and <112> axes. Near coincidence site and 0-lattice theories indicated that the {111}‖{111} grain-boundary interconnection usually corresponded to higher planar coincidence site densities and definite dislocation structures.

The segregation along deformation-induced interfaces in the matrix of Mg-2Y-1at%Zn alloy, compressed at 473K was studied[291], showing that kink, twin-like and tilt boundaries within matrix grains contained concentrations of zinc and yttrium. Particularly in the case of various tilt boundaries, the segregation could be explained in terms of 0-lattice theory or partial dislocation effects. The segregation at kink and twin-like boundaries was closely linked to associated partial dislocations.

The structures of low-angle and high-angle (001), (110) and (111) tilt boundaries in thin gold bicrystalline films, when examined[292] using transmission electron microscopy, revealed periodic strain contrast patterns. The grain-boundary structure was in general agreement with 0-lattice theory, although deviations occurred in the case of low-angle (111) tilt boundaries, and at certain special boundary inclinations.

The Burgers vectors of misfit dislocations in the $Ti_{40}Ni_{40}Nb_{20}$ eutectic were analyzed[293] using 0-lattice theory. This shape memory alloy consists of a B2-structured TiNi α-phase and a body-centerd cubic Nb-rich β-phase. The eutectic morphology was lamellar, with {110}α‖{110}β. The orientational relationship was: [100]α‖[100]β [010]α‖[010]β [001]α‖[001]β, so the equivalent directions and planes were thus exactly parallel.

Four near coincidence-site orientational relationships, between MgO precipitates and a copper matrix, were identified: Σ41 [110], Σ13 [111], Σ29 [100], Σ35 [112]. The size of the MgO precipitates which exhibited these orientations ranged from 0.5 to 1μm. Possible dislocation networks for the relationships were analyzed by using 0-lattice theory. The common occurrence of these special orientations implied[294] that they were the most favoured ones for the precipitation and coarsening of MgO particles.

Table 14. Predicted interface structures for graphene on (111) metal substrates

Substrate	a	ε(%)	ε(%)*	S_s*	S_g*
Pt	3.98	12.55	0.98	√3 x √3	2 x 2
Cu	3.62	3.92	-3.92	1 x 1	1 x 1
Au	4.17	16.45	-3.53	√3 x √3	2 x 2
Pd	3.95	11.71	1.94	√3 x √3	2 x 2
Ni	3.52	0.93	-0.93	1 x 1	1 x 1
Al	4.07	14.37	-1.12	√3 x √3	2 x 2
Ag	4.14	15.82	-2.79	√3 x √3	2 x 2

*computed; ε, misfit; S_s, substrate super-cell; S_g, graphene super-cell

The common super-cell found is generally a compromise between one containing a small number of atoms and one exhibiting minimal strain. Any program therefore performs an intensive unsophisticated search for a common super-cell where one material is forced to match another. The possible structures are first sorted according to the level of strain. In the case of graphene on the (111) plane of iridium, for example, this is quite easy and the common super-cell consists of a moiré pattern of 10 x 10 graphene unit cells overlying 9 x 9 (111) iridium unit cells. The program takes account only of the geometry of the super-cell and the positions of the contained atoms, and uses first-principles density functional approaches with no empirical parameters. The method also incidentally furnishes the band-structure of the super-cell.

A search program has been used to determine the lattice-matching of graphene, with its lattice parameter of 2.46Å, against the (111) surfaces of platinum, palladium, copper, silver, nickel, gold and aluminium. A 1:1 match between the primitive cells of graphene and the substrate results in lattice-mismatches which exceed 10%, except in the case of copper and nickel (table 14).

Lattice-Matching of Organic Materials

Unlike the materials mentioned above, organic molecules tend to be large, irregularly-shaped and of low symmetry: very different to their epitaxial substrates. The familiar technique of aligning primitive lattice-vectors in the overlayer with primitive lattice-vectors in the substrate, and optimizing the misfit, will usually not work. There nevertheless exist models and geometrical matching strategies which can be used to model epitaxy. These are based upon an assumed correspondence between lattice-position and potential energy surfaces. This permits the interface to be described by superposing plane waves.

It was found relatively recently[295] that there exist adsorbate/substrate combinations having definite mutual orientations, but for which the epitaxy cannot be explained using rigid-lattice concepts. It had been proposed that small deviations (so-called static distortion waves) of the atomic positions from ideal lattice-points were responsible for the orientational epitaxy observed in such cases. In the case of hexa-peri-hexabenzocoronene on graphite, low-energy electron diffraction and scanning tunnelling microscopy had provided direct experimental evidence for static distortion waves. They were seen as sub-Ångström wave-like displacements away from an ideal adsorbate lattice which was incommensurate with the graphite. Due to the flexibility of the adsorbate layer, energy was gained by straining the intermolecular bonds and the resultant total energy was a minimum for the observed orientational epitaxy. This renders more complicated the design of molecular films. The models of epitaxy which were thought to describe these films were based upon 7 parameters.

These were the lattice parameters of the substrate (a_1, a_2, α), the lattice parameters of the overlayer (b_1, b_2, β) and the azimuthal angle, θ, between the lattice vectors, a_1 and b_1. The substrate and overlayer lattice vectors for a given θ were then related via a four-element transformation matrix whose coefficients described the relationship between the lattice vectors of the substrate and overlayer. The matrix alone could be used to characterize epitaxial configurations. The magnitudes of the matrix elements depended upon the primitive lattice vectors which were chosen to describe the primitive unit cell. Because all of the latter cells had to have equal areas, the matrix determinants were identical.

An overlayer unit cell could always be constructed so as to have a reciprocal basis vector, b_1^*, which coincided with a substrate reciprocal basis vector, a_1^*, and ensured a commensurate or coincidence relationship. The vectors were related by $b_1^* = m a_1^*$, with m an integer. When all of the matrix elements were integers, the overlayer was commensurate with the substrate. When the matrix elements were all rational numbers, but with only 2 integers confined to one column, each overlayer lattice-point rested atop at least one primitive lattice-line of the substrate. This was termed a point-online coincidence. If the super-cell positions which coincided with substrate lattice points were assumed to be energetically preferred, this condition implied that overlayer lattice points which were located on the perimeter were less favorable. The behaviour of the overlayer was governed by whether it was energetically preferable for the overlayer to rotate as a whole by some angle in order to achieve coincidence, or for the overlayer to become strained in order to become commensurate.

If the matrix contained 4 rational numbers, but integers were present in columns, this corresponded to a sort of point-online coincidence with every other row of overlayer lattice points being situated between rows of substrate lattice points. Line-on-line coincidence has been found in organic/organic heterolayers where the overlayer molecules are here situated on parallel equally-spaced lines; these lines not being limited to being primitive substrate lattice lines. The observed epitaxial ordering corresponds to a minimum potential energy. Another complication to be recognised is that the lowest-energy state can exhibit a clearly non-symmetrical relative orientation angle.

A combination of low-energy electron diffraction, scanning tunnelling microscopy and density functional theory has lately revealed the existence of static distortion waves in overlayers of certain organic molecules. Although other overlayers had been suspected to obey the static distortion wave model, the existence of a sinusoidal incommensurate state existing across an entire substrate was difficult to detect. In the case in question, the static distortion wave structure was detected in a moiré pattern of triangles where average displacements of only 0.52Å pointed away from the corners of the triangles. The distortion was attributed to a gain in molecule-substrate energy which was associated with strain in the slightly elastic overlayer. The distortion leads to non-integer elements, in the matrices for room-temperature and low-temperature phases, which could be easily taken to signify a commensurate arrangement.

Thin-film epitaxy involves the superposition of rigid two-dimensional lattices having a well-defined orientation governed by commensurate lattice spacings[296]. The organization of organic molecules on substrates is complicated however by static distortion waves which impose minute shifts of atomic positions away from substrate lattice points. This can lead to orientations of a molecular film which cannot be handled by standard models. Attention has therefore been focussed on configurations which reflect the balance between intermolecular interactions within a molecular film and molecule-substrate interactions. Geometrical models thus need to take account of energetic factors.

The epitaxial growth of organic/organic multilayer structures is a relatively neglected subject[297]. One comparison of experiment and simulation showed however that highly anisotropic needle-like para-hexaphenyl crystallites can act as an organic template and epitaxial overgrowth by 2,2':6',2''-ternaphthalene leads to high molecular order and optical anisotropy of the ternaphthalene crystallites. Surface corrugations formed by the para-hexaphenyl template cause a parallel molecular alignment and herringbone stacking sequence of the ternaphthalene. It was concluded that lattice-matching plays a minor role, and that adoption of the molecular stacking is due directly to optimization of the adsorption energy.

Miscellaneous Lattice-Matchings

The similar van der Waals epitaxy of three-dimensional CdS thin films on monocrystalline graphene/Cu(111)/spinel(111) and monocrystalline graphene/SiO$_2$/Si substrates was studied[298]. X-ray and electron back-scattering diffraction pole-figures showed that the CdS films had a Wurtzite structure, with a weak epitaxy on graphene, together with a fiber-texture background. The epitaxial alignment between CdS and graphene involved an unusual non-parallel epitaxial relationship with a 30° rotation between the unit vectors of CdS and graphene. A geometrical model, based upon the minimization of superlattice areal mismatch, was used to calculate possible interface lattice arrangements. The 30° rotation between CdS and graphene was indeed the most probable interface epitaxial lattice alignment .

Films of Ge$_{1-x}$Sn$_x$ were grown on monocrystalline (111) and (100) CaF$_2$ substrates in order to study[299] the effect of tin upon germanium crystallization. Single-orientation Ge$_{1-x}$Sn$_x$ (111) films grew on (111) CaF$_2$ substrates at various temperatures, but a temperature-dependent superposition of (111) and (100) orientations occurred in the case of films which were grown on (100) CaF$_2$ at above 250C. This had been successfully predicted by using a superlattice area-matching method.

An increased ionic conductivity at the nanoscale planar interfaces of the CaF$_2$‖BaF$_2$ system was successfully modelled[300] by using molecular dynamics simulations. A criterion was developed for the choice of simulation cells which contained any given lattice-mismatched interfaces while maintaining periodic boundary conditions. Both of the fluorides were cubic, so that changes in the cell dimensions caused simultaneous dimensional changes in all 3 crystallographic dimensions. A volume change of 1% caused a structural energy-change of only a few kJ/mol, which was acceptable in building a super-cell of the CaF$_2$‖BaF$_2$ planar heterostructure, in which any mismatch would be removed by stretching or compressing one or both fluorides. The minimum lattice dimensions that gave less than 1% of super-cell lattice mismatch for the (001)‖(001) and (111)‖(111) interfaces were about 4 and 6nm, respectively. Thus if a nanocomposite of these fluorides were to be used to form a superionic conductor, one set of favourable dimensions for individual crystallite would be 4 or 6nm. A random selection of crystallite dimensions could cause greater superlattice mismatches, drive a recrystallization leading to solid-solution formation or amorphization and eliminate most of the defects that would increase the mobility and hence the ionic conductivity. The relative-to-bulk ionic conductivity increase at the (111)CaF$_2$‖(111)BaF$_2$ interface was qualitatively reproduced. A factor-of-7.6 higher conductivity was predicted for the (001)CaF$_2$‖(001)BaF$_2$ interface.

In the case of photovoltaic technology, it is also important to match electron and hole extraction layers in order to make an efficient device. It has recently been suggested[301] that enargite (Cu_3AsS_4) and bournonite ($CuPbSbS_3$) minerals would be suitable materials, given that they are chemically stable and possess the required opto-electronic properties to be used as the absorber layer in a thin-film photovoltaic device. In these compounds, spontaneous lattice polarization, with internal electric fields, could promote increased carrier separation and new photophysical effects. The ionization potentials were calculated for non-polar surface terminations, and suitable partners were sought for the formation of solar-cell heterojunctions by matching electronic band edges to a short-list of viable electrical materials. These candidates were then further whittled down by matching lattice constants with a view to minimising strain and ensuring epitaxy. This screening identified somewhat surprising junction partners which included SnS_2, ZnTe, WO_3 and Bi_2O_3.

The microstructures of hexa-peri-hexabenzocoronene ($C_{42}H_{18}$) thin films, deposited onto inert substrates of similar surface energy, were studied[302]. This showed that it formed polycrystalline films on SiO_2, with the molecules oriented in an upright position and exhibiting the bulk structure. On the other hand, films which were deposited onto highly-oriented pyrolytic graphite exhibited a novel substrate-induced polymorphic form in which all of the molecules were recumbent, with a planar π-stacking. The formation of this phase depended very much upon the coherence of the underlying graphite lattice. That is, when grown onto defective highly-oriented pyrolytic graphite it again exhibited the same orientation and phase as when deposited onto SiO_2. This demonstrates that the resultant film structure and morphology are not determined only by the adsorption energy, but also by symmetry-matching and lattice-matching between the substrate and ad-molecules. It also reveals that highly coherent but weakly-interacting substrates can be useful in creating new polymorphs having molecular arrangements which are very different to that of the bulk structure.

References

[1] Rosciano, F., Pescarmona, P.P., Houthoofd, K., Persoons, A., Bottke, P., Wilkening, M., Physical Chemistry Chemical Physics, 15[16] 2013, 6107-6112. https://doi.org/10.1039/c3cp50803j

[2] Zou, M., Yang, F., Wen, K., Lv, W., Waqas, M., He, W., International Journal of Hydrogen Energy, 41[47] 2016, 22254-22259. https://doi.org/10.1016/j.ijhydene.2016.10.013

[3] Liao, Z., Green, R.J., Gauquelin, N., Macke, S., Li, L., Gonnissen, J., Sutarto, R.,

Houwman, E.P., Zhong, Z., Van Aert, S., Verbeeck, J., Sawatzky, G.A., Huijben, M., Koster, G., Rijnders, G., Advanced Functional Materials, 26[36] 2016, 6627-6634. https://doi.org/10.1002/adfm.201602155

[4] Butler, K.T., Hendon, C.H., Walsh, A., Faraday Discussions, 201, 2017, 207-219. https://doi.org/10.1039/C7FD00019G

[5] Davenport, H., The Mathematical Gazette, 31[296] 1947, 206-210. https://doi.org/10.2307/3608159

[6] Friedel, G., Lecons de Cristallographie, Blenchard, 1926.

[7] Balluffi, R.W., Brokman, A., King, A.H., Acta Metallurgica, 30[8] 1982, 1453-1470. https://doi.org/10.1016/0001-6160(82)90166-3

[8] Morita, K., Tsurekawa, S., Nakashima, H., Yoshinaga, H., Materials Science Forum, 204-206[1] 1996, 239-244 https://doi.org/10.4028/www.scientific.net/MSF.204-206.239

[9] Miyano, N., Ameyama, K., Journal of the Japan Institute of Metals, 64[1] 2000, 42-49. https://doi.org/10.2320/jinstmet1952.64.1_42

[10] King, A.H., Singh, A., Journal of Physics and Chemistry of Solids, 55[10] 1994, 1023-1033. https://doi.org/10.1016/0022-3697(94)90122-8

[11] Shiue, Y.R., Phillips, D.S., Phillips, D.S., Philosophical Magazine A, 50[5] 1984, 677-702. https://doi.org/10.1080/01418618408237527

[12] Fukutomi, Hiroshi, Tanaka, Mutsuto, Nippon Kinzoku Gakkaisi, 49[8] 1985, 607-613.

[13] Zhu, Y., Suenaga, M., Philosophical Magazine A, 66[3] 1992, 457-471. https://doi.org/10.1080/01418619208201569

[14] Chen, L.Q., Kalonji, G., Philosophical Magazine A, 66[1] 1992, 11-26. https://doi.org/10.1080/01418619208201510

[15] Thibault, J., Putaux, J.L., Jacques, A., George, A., Michaud, H.M., Baillin, X., Materials Science and Engineering A, 164[1-2] 1993, 93-100. https://doi.org/10.1016/0921-5093(93)90646-V

[16] Mou, Y., Metallurgical and Materials Transactions A, 25[9] 1994, 1905-1915. https://doi.org/10.1007/BF02649038

[17] Morniroli, J.P., Cherns, D., Ultramicroscopy, 62[1-2] 1996, 53-63. https://doi.org/10.1016/0304-3991(95)00087-9

[18] Antoniades, I.P., Bleris, G.L., Philosophical Magazine A, 80[12] 2000, 2871-2897. https://doi.org/10.1080/01418610008223900

[19] Poulat, S., Thibault, J., Priester, L., Interface Science, 8[1] 2000, 5-15. https://doi.org/10.1023/A:1008729517242

[20] Gertsman, V.Y., Acta Crystallographica A, 58[2] 2002, 155-161. https://doi.org/10.1107/S0108767301020219

[21] Hu, J.R., Chang, S.C., Chen, F.R., Kai, J.J., Materials Chemistry and Physics, 74[3] 2002, 313-319. https://doi.org/10.1016/S0254-0584(01)00484-9

[22] Couzinié, J.P., Decamps, B., Priester, L., Philosophical Magazine Letters, 83[12] 2003, 721-731. https://doi.org/10.1080/09500830310001614522

[23] Ikuhara, Y., Nishimura, H., Nakamura, A., Matsunaga, K., Yamamoto, T., Peter, K., Lagerlöf, D., Journal of the American Ceramic Society, 86[4] 2003, 595-602. https://doi.org/10.1111/j.1151-2916.2003.tb03346.x

[24] Décamps, B., Priester, L., Thibault, J., Advanced Engineering Materials, 6[10] 2004, 814-818. https://doi.org/10.1002/adem.200400086

[25] Hyde, B., Farkas, D., Caturla, M.J., Philosophical Magazine, 85[32] 2005, 3795-3807. https://doi.org/10.1080/14786430500256342

[26] Kizuka, T., Japanese Journal of Applied Physics - 1, 46[11] 2007, 7396-7398. https://doi.org/10.1143/JJAP.46.7396

[27] Mompiou, F., Legros, M., Caillard, D., Materials Research Society Symposium Proceedings, 1086, 2008, 19-24. https://doi.org/10.1557/PROC-1086-U09-04

[28] Sheikh-Ali, A.D., Acta Materialia, 58[19] 2010, 6249-6255. https://doi.org/10.1016/j.actamat.2010.07.043

[29] Mitsuma, T., Tohei, T., Shibata, N., Mizoguchi, T., Yamamoto, T., Ikuhara, Y., Journal of Materials Science, 46[12] 2011, 4162-4168. https://doi.org/10.1007/s10853-011-5266-5

[30] Tsuru, T., Kaji, Y., Shibutani, Y., Journal of Applied Physics, 110[7] 2011, 073520. https://doi.org/10.1063/1.3651384

[31] Wan, L., Li, J., Modelling and Simulation in Materials Science and Engineering, 21[5] 2013, 055013. https://doi.org/10.1088/0965-0393/21/5/055013

[32] Saito, M., Wang, Z., Tsukimoto, S., Ikuhara, Y., Journal of Materials Science, 48[16] 2013, 5470-5474. https://doi.org/10.1007/s10853-013-7340-7

[33] Wan, L., Han, W., Chen, K., Scientific Reports, 5, 2015, 13441. https://doi.org/10.1038/srep13441

[34] Ishikawa, R., Lugg, N.R., Inoue, K., Sawada, H., Taniguchi, T., Shibata, N., Ikuhara, Y., Scientific Reports, 6, 2016, 21273. https://doi.org/10.1038/srep21273

[35] Sharma, N.K., Shekhar, S., Philosophical Magazine, 97[23] 2017, 2004-2017. https://doi.org/10.1080/14786435.2017.1322730

[36] Annevelink, E., Ertekin, E., Johnson, H.T., Acta Materialia, 166, 2019, 67-74. https://doi.org/10.1016/j.actamat.2018.12.030

[37] Potin, V., Ruterana, P., Nouet, G., Pond, R., Physical Review B, 61[8] 2000, 5587-5599. https://doi.org/10.1103/PhysRevB.61.5587

[38] Ma, X., Guo, X., Fu, M., Intermetallics, 98, 2018, 11-17. https://doi.org/10.1016/j.intermet.2018.04.007

[39] Deng, H., Dickey, E.C., Paderno, Y., Paderno, V., Filippov, V., Journal of the American Ceramic Society, 90[8] 2007, 2603-2609. https://doi.org/10.1111/j.1551-2916.2007.01812.x

[40] Yu, R., He, L.L., Guo, J.T., Ye, H.Q., Lupinc, V., Acta Materialia, 48[14] 2000, 3701-3710. https://doi.org/10.1016/S1359-6454(00)00167-1

[41] Winkelman, G.B., Raviprasad, K., Muddle, B.C., Acta Materialia, 55[9] 2007, 3213-3228. https://doi.org/10.1016/j.actamat.2007.01.011

[42] Zhang, W.Z., Weatherly, G.C., Progress in Materials Science, 50[2] 2005, 181-292. https://doi.org/10.1016/j.pmatsci.2004.04.002

[43] Xu, W.S., Yang, X.P., Zhang, W.Z., Acta Metallurgica Sinica, 31[2] 2018, 113-126. https://doi.org/10.1007/s40195-017-0693-1

[44] Bollmann, W., Surface Science, 31, 1972, 1-11. https://doi.org/10.1016/0039-6028(72)90250-6

[45] Bonnet, R., Durand, F., Materials Research Bulletin, 7[10] 1972, 1045-1059. https://doi.org/10.1016/0025-5408(72)90157-2

[46] Smith, D.A., Pond, R.C., International Metals Reviews, 21[1] 1976, 61-74. https://doi.org/10.1179/imtr.1976.21.1.61

[47] Sadananda, K., Marcinkowski, M.J., Journal of Applied Physics, 45[4] 1974, 1521-1532. https://doi.org/10.1063/1.1663454

[48] Sadananda, K., Marcinkowski, M.J., Scripta Metallurgica, 7[6] 1973, 557-563.

https://doi.org/10.1016/0036-9748(73)90215-9

[49] Bollmann, W., Scripta Metallurgica, 7[6] 1973, 565-568.
https://doi.org/10.1016/0036-9748(73)90216-0

[50] Mourey, M., Dabosi, F., Materials Research Bulletin, 9[4] 1974, 379-389.
https://doi.org/10.1016/0025-5408(74)90205-0

[51] Mou, Y., Aaronson, H.I., Acta Metallurgica et Materialia, 42[6] 1994, 2133-2144.
https://doi.org/10.1016/0956-7151(94)90038-8

[52] Bollmann, W., Poerry, A.J., Philosophical Magazine, 20[163] 1969, 33-50.
https://doi.org/10.1080/14786436908228534

[53] Warrington, D.H., Bollmann, W., Philosophical Magazine, 25[5] 1972, 1195-1199.
https://doi.org/10.1080/14786437208226861

[54] Shin, K., King, A.H., Materials Science and Engineering A, 113, 1989, 121-127.
https://doi.org/10.1016/0921-5093(89)90298-0

[55] Rodríguez, A.G., Casajuana, D.R., Revista Mexicana de Fisica, 53[2] 2007, 139-143.

[56] Zhang, W.Z., Yang, X.P., Journal of Materials Science, 46[12] 2011, 4135-4156.
https://doi.org/10.1007/s10853-011-5431-x

[57] Chen, H., Yang, Z., Acta Metallurgica Sinica, 43[7] 2007, 710-712.

[58] Ecob, R.C., Physica Status Solidi A, 71[2] 1982, 399-407.
https://doi.org/10.1002/pssa.2210710214

[59] Zhang, J.Y., Gao, Y., Wang, Y., Zhang, W.Z., Acta Materialia, 165, 2019, 508-519.
https://doi.org/10.1016/j.actamat.2018.12.005

[60] Dahmen, U., Scripta Metallurgica, 15[1] 1981, 77-81. https://doi.org/10.1016/0036-9748(81)90140-X

[61] Kato, M., Materials Transactions, JIM, 33[2] 1992, 89-96.
https://doi.org/10.2320/matertrans1989.33.89

[62] Kato, M., Materials Science and Engineering A, 146[1-2] 1991, 205-216.
https://doi.org/10.1016/0921-5093(91)90278-U

[63] Yuryev, D.V., Demkowicz, M.J., Applied Physics Letters, 105[22] 2014, 221601.
https://doi.org/10.1063/1.4902888

[64] Zhang, W.Z., Metallurgical and Materials Transactions A, 44[10] 2013, 4513-4531.
https://doi.org/10.1007/s11661-013-1689-8

[65] Romeu, D., Gómez-Rodríguez, A., Acta Crystallographica A, 62[5] 2006, 411-412. https://doi.org/10.1107/S0108767306025293

[66] Zhang, W.Z., Qiu, D., Yang, X.P., Ye, F., Metallurgical and Materials Transactions A, 37[3] 2006, 911-927. https://doi.org/10.1007/s11661-006-0065-3

[67] Tsurekawa, S., Morita, K., Nakashima, H., Yoshinaga, H., Materials Transactions, 38[5] 1997, 393-400. https://doi.org/10.2320/matertrans1989.38.393

[68] Goux, C., Canadian Metallurgical Quarterly, 13[1] 1974, 9-31. https://doi.org/10.1179/000844374795595262

[69] Lojkowski, W., Materials Research Society Symposium - Proceedings, 357, 1995, 407-412.

[70] Sangghaleh, A., Demkowicz, M.J., Computational Materials Science, 145, 2018, 35-47. https://doi.org/10.1016/j.commatsci.2017.12.025

[71] Bollmann, W., Michaut, B., Sainfort, G., Physica Status Solidi A, 13[2] 1972, 637-649. https://doi.org/10.1002/pssa.2210130236

[72] Luo, C.P., Weatherly, G.C., Acta Metallurgica, 35[8] 1987, 1963-1972. https://doi.org/10.1016/0001-6160(87)90025-3

[73] Chen, Fu-Rong, King, A.H., Metallurgical Transactions. A, 19[9] 1988, 2359-2363. https://doi.org/10.1007/BF02645061

[74] Cherkashin, N., Kononchuk, O., Hÿtch, M., Solid State Phenomena, 178-179, 2011, 489-494. https://doi.org/10.4028/www.scientific.net/SSP.178-179.489

[75] Cherkashin, N., Kononchuk, O., Reboh, S., Hÿtch, M., Acta Materialia, 60[3] 2012, 1161-1173. https://doi.org/10.1016/j.actamat.2011.10.054

[76] Vaudin, M.D., Rühle, M., Sass, S.L., Acta Metallurgica, 31[7] 1983, 1109-1116. https://doi.org/10.1016/0001-6160(83)90206-7

[77] Eastman, J., Schmueckle, F., Vaudin, M.D., Sass, S.L., Advances in Ceramics, 10, 1984, 324-346.

[78] Zhu, Y., Philosophical Magazine A, 69[4] 1994, 717-728. https://doi.org/10.1080/01418619408242513

[79] Shashkov, D.A., Chisholm, M.F., Seidman, D.N., Acta Materialia, 47[15] 1999, 3939-3951. https://doi.org/10.1016/S1359-6454(99)00255-4

[80] Ramamurthy, S., Carter, C.B., Physica Status Solidi A, 166[1] 1998, 37-55. https://doi.org/10.1002/(SICI)1521-396X(199803)166:1<37::AID-

PSSA37>3.0.CO;2-W

[81] Bartholomeusz, B.J., Lu, T.M., Rajan, K., Journal of Electronic Materials, 20[7] 1991, 759-765. https://doi.org/10.1007/BF02665962

[82] Ramamoorthy, K., Sanjeeviraja, C., Jayachandran, M., Sankaranarayanan, K., Misra, P., Kukreja, L.M., Materials Chemistry and Physics, 84[1] 2004, 14-19. https://doi.org/10.1016/j.matchemphys.2003.09.001

[83] Zur, A., McGill, T.C., Journal of Applied Physics. 55[2] 1984, 378-386. https://doi.org/10.1063/1.333084

[84] Büschel, M., Tempel, A., Zehe, A., Crystal Research and Technology, 26[2] 1991, 211-215. https://doi.org/10.1002/crat.2170260215

[85] Myers, T.H., Lo, Y., Bicknell, R.N., Schetzina, J.F., Applied Physics Letters, 42[3], 1983, 247-248. https://doi.org/10.1063/1.93903

[86] Raclariu, A.M., Deshpande, S., Bruggemann, J., Zhuge, W., Yu, T.H., Ratsch, C., Shankar, S., Computational Materials Science, 108, 2015, 88-93. https://doi.org/10.1016/j.commatsci.2015.05.023

[87] Mathew, K., Singh, A.K., Gabriel, J.J., Choudhary, K., Sinnott, S.B., Davydov, A.V., Tavazza, F., Hennig, R.G., Computational Materials Science, 122, 2016, 183-190. https://doi.org/10.1016/j.commatsci.2016.05.020

[88] Efimov, A.N., Lebedev, A.O., Thin Solid Films, 260[1] 1995, 111-117. https://doi.org/10.1016/0040-6090(94)06425-3

[89] May, B.J., Anderson, P.M., Myers, R.C., Journal of Crystal Growth, 459, 2017, 209-214. https://doi.org/10.1016/j.jcrysgro.2016.11.042

[90] Runnels, B., Beyerlein, I.J., Conti, S., Ortiz, M., Journal of the Mechanics and Physics of Solids, 89, 2016, 174-193. https://doi.org/10.1016/j.jmps.2016.01.008

[91] Jelver, L., Larsen, P.M., Stradi, D., Stokbro, K., Jacobsen, K.W., Physical Review B, 96, 2017, 085306. https://doi.org/10.1103/PhysRevB.96.085306

[92] Efimov, A.N., Lebedev, A.O., Surface Science, 344[3] 1995, 276-282. https://doi.org/10.1016/0039-6028(95)00843-8

[93] Lazić, P., Computer Physics Communications, 197, 2015, 324-334. https://doi.org/10.1016/j.cpc.2015.08.038

[94] Kawahara, K., Arafune, R., Kawai, M., Takagi, N., e-Journal of Surface Science and Nanotechnology, 13, 2015, 361-365. https://doi.org/10.1380/ejssnt.2015.361

[95] Zur, A., McGill, T.C., Nicolet, M.A., Journal of Applied Physics, 57[2] 1985, 600-603. https://doi.org/10.1063/1.334743

[96] Croke, E.T., Hauenstein, R.J., Nieh, C.W., McGill, T.C., Journal of Electronic Materials, 18[6] 1989, 757-761. https://doi.org/10.1007/BF02657529

[97] Croke, E.T., Hauenstein, R.J., McGill, T.C., Applied Physics Letters, 53[6] 1988, 514-516. https://doi.org/10.1063/1.100621

[98] Dakshinamurthy, S., Rajan, K., JOM, 46[3] 1994, 52-53. https://doi.org/10.1007/BF03220652

[99] Bulle-Lieuwma, C.W.T., Van Ommen, A.H., Hornstra, J., Aussems, C.N.A.M., Journal of Applied Physics, 71[5] 1992, 2211-2224. https://doi.org/10.1063/1.351119

[100] Lin, W.T., Wu, K.C., Pan, F.M., Thin Solid Films, 215[2] 1992, 184-187. https://doi.org/10.1016/0040-6090(92)90435-E

[101] Kang, T.S., Je, J.H., Kim, G.B., Baik, H.K., Lee, S.M., Journal of Vacuum Science and Technology B, 18[4] 2000, 1953-1956. https://doi.org/10.1116/1.1305275

[102] Smeets, D., Vantomme, A., De Keyser, K., Detavernier, C., Lavoie, C., Journal of Applied Physics, 103[6] 2008, 063506. https://doi.org/10.1063/1.2888554

[103] Shiau, F.Y., Cheng, H.C., Chen, L.J., Journal of Applied Physics, 59[8] 1986, 2784-2787. https://doi.org/10.1063/1.336990

[104] Heck, C., Kusaka, M., Hirai, M., Iwami, M., Yokota, Y., Thin Solid Films, 281-282[1-2] 1996, 94-97. https://doi.org/10.1016/0040-6090(96)08583-5

[105] Filonenko, O., Mogilatenko, A., Hortenbach, H., Allenstein, F., Beddies, G., Hinneberg, H.J., Journal of Crystal Growth, 262[1-4] 2004, 281-286. https://doi.org/10.1016/j.jcrysgro.2003.10.054

[106] Van An, F., Bulenkov, N.A., Andreeva, A.V., Physica Status Solidi A, 88[2] 1985, 429-441. https://doi.org/10.1002/pssa.2210880205

[107] Dascalu, M., Cesura, F., Levi, G., Diéguez, O., Kohn, A., Goldfarb, I., Applied Surface Science, 476, 2019, 189-197. https://doi.org/10.1016/j.apsusc.2019.01.079

[108] Geib, K.M., Mahan, J.E., Long, R.G., Nathan, M., Bai, G., Journal of Applied Physics, 70[3] 1991, 1730-1736. https://doi.org/10.1063/1.349543

[109] Mahan, J.E., Geib, K.M., Robinson, G.Y., Long, R.G., Xinghua, Y., Bai, G., Nicolet, M.A., Nathan, M., Applied Physics Letters, 56[21] 1990, 2126-2128. https://doi.org/10.1063/1.103235

Materials Research Forum LLC
https://doi.org/10.21741/9781644900475

[110] Catana, A., Schmid, P.E., Heintze, M., Lévy, F., Stadelmann, P., Bonnet, R., Journal of Applied Physics, 67[4] 1990, 1820-1825. https://doi.org/10.1063/1.345609

[111] Xie, W., Lucking, M., Chen, L., Bhat, I., Wang, G.C., Lu, T.M., Zhang, S., Crystal Growth and Design, 16[4] 2016, 2328-2334. https://doi.org/10.1021/acs.cgd.6b00118

[112] Kim, K.H., Lee, J.J., Seo, D.J., Choi, C.K., Hong, S.R., Koh, J.D., Kim, S.C., Lee, J.Y., Nicolet, M.A., Journal of Applied Physics, 71[8] 1992, 3812-3815. https://doi.org/10.1063/1.350895

[113] Huang, J.Y., Wu, S.T., Japanese Journal of Applied Physics - 1, 38[6A] 1999, 3660-3663. https://doi.org/10.1143/JJAP.38.3660

[114] He, Z., Stevens, M., Smith, D.J., Bennett, P.A., Surface Science, 524[1-3] 2003, 148-156. https://doi.org/10.1016/S0039-6028(02)02506-2

[115] Goldfarb, I., Grossman, S., Cohen-Taguri, G., Applied Surface Science, 252[15] 2006, 5355-5360. https://doi.org/10.1016/j.apsusc.2005.12.026

[116] Lin, W.T., Chen, L.J., Journal of Applied Physics, 59[5] 1986, 1518-1524. https://doi.org/10.1063/1.336458

[117] Chu, J.J., Chen, L.J., Tu, K.N., Journal of Applied Physics, 63[4] 1988, 1163-1167. https://doi.org/10.1063/1.340024

[118] Mahan, J.E., Geib, K.M., Robinson, G.Y., Long, R.G., Xinghua, Y., Bai, G., Nicolet, M.A., Nathan, M., Applied Physics Letters, 56[24] 1990, 2439-2441. https://doi.org/10.1063/1.103251

[119] Lee, Y.K., Fujimura, N., Ito, T., Itoh, N., Nanostructured Materials, 2[6] 1993, 603-614. https://doi.org/10.1016/0965-9773(93)90034-9

[120] Lin, W.T., Chen, L.J., Journal of Applied Physics, 59[10] 1986, 3481-3488. https://doi.org/10.1063/1.336818

[121] Chen, J.F., Chen, L.J., Thin Solid Films, 261[1-2] 1995, 107-114. https://doi.org/10.1016/S0040-6090(95)06520-2

[122] Kawarada, H., Ishida, M., Nakanishi, J., Ohdomari, I., Horiuchi, S., Philosophical Magazine A, 54[5] 1986, 729-741. https://doi.org/10.1080/01418618608244029

[123] Konuma, K., Utsumi, H., Journal of Applied Physics, 76[4] 1994, 2181-2184. https://doi.org/10.1063/1.357631

[124] Kavanagh, K.L., Reuter, M.C., Tromp, R.M., Journal of Crystal Growth, 173[3-4]

1997, 393-401. https://doi.org/10.1016/S0022-0248(96)01047-0

[125] Du, Y., Chen, K.H., Schuster, J.C., Perring, L., Huang, B.Y., Yuan, Z.H., Gachon, J.C., Zeitschrift für Metallkunde, 92[4] 2001, 323-327.

[126] Chang, Y.S., Chou, M.L., Journal of Applied Physics, 68[5] 1990, 2411-2414. https://doi.org/10.1063/1.346500

[127] Chang, Y.S., Chu, J.J., Materials Letters, 5[3] 1987, 67-71. https://doi.org/10.1016/0167-577X(87)90077-2

[128] Chang, C.S., Nieh, C.W., Chu, J.J., Chen, L.J., Thin Solid Films, 161, 1988, 263-271. https://doi.org/10.1016/0040-6090(88)90258-1

[129] Chen, Q., Xie, Q., Physics Procedia, 11, 2011, 134-137. https://doi.org/10.1016/j.phpro.2011.01.022

[130] Xiao, Q., Xie, Q., Shen, X., Zhang, J., Yu, Z., Zhao, K., Applied Surface Science, 257[17] 2011, 7800-7804. https://doi.org/10.1016/j.apsusc.2011.04.032

[131] Chen, J.C., Shen, G.H., Chen, L.J., Journal of Applied Physics, 84[11] 1998, 6083-6087. https://doi.org/10.1063/1.368920

[132] Liu, B.Z., Nogami, J., Nanotechnology, 14[8] 2003, 873-877. https://doi.org/10.1088/0957-4484/14/8/306

[133] Lee, Y.K., Fujimura, N., Ito, T., Itoh, N., Journal of Crystal Growth, 134[3-4] 1993, 247-254. https://doi.org/10.1016/0022-0248(93)90133-H

[134] Lee, Y.K., Lee, M.S., Lee, J.S., Journal of Crystal Growth, 244[3-4] 2002, 305-312. https://doi.org/10.1016/S0022-0248(02)01695-0

[135] Frangis, N., Van Landuyt, J., Kaltsas, G., Travlos, A., Nassiopoulos, A.G., Journal of Crystal Growth, 172[1-2] 1997, 175-182. https://doi.org/10.1016/S0022-0248(96)00745-2

[136] Yang, Y., Abelson, J.R., Journal of Crystal Growth, 310[13] 2008, 3197-3202. https://doi.org/10.1016/j.jcrysgro.2008.03.035

[137] Li, B.Q., Zuo, J.M., Surface Science, 520[1-2] 2002, 7-17. https://doi.org/10.1016/S0039-6028(02)02313-0

[138] Jin, H.S., Park, K.H., Yapsir, A.S., Wang, G.C., Lu, T.M., Luo, L., Gibson, W.M., Yamada, I., Takagi, T., Nuclear Instruments and Methods in Physics Research B, 40-41[2] 1989, 817-822. https://doi.org/10.1016/0168-583X(89)90485-0

[139] Jin, H.S., Yapsir, A.S., Lu, T.M., Gibson, W.M., Yamada, I., Takagi, T., Applied

Physics Letters, 50[16] 1987, 1062-1064. https://doi.org/10.1063/1.97970

[140] Legoues, F.K., Liehr, M., Renier, M., Krakow, W., Philosophical Magazine B, 57[2] 1988, 179-189. https://doi.org/10.1080/13642818808201613

[141] Goswami, D.K., Bhattacharjee, K., Satpati, B., Roy, S., Kuri, G., Satyam, P.V., Dev, B.N., Applied Surface Science, 253[23] 2007, 9142-9147. https://doi.org/10.1016/j.apsusc.2007.05.051

[142] Nason, T.C., You, L., Lu, T.M., Journal of Applied Physics, 72[2] 1992, 466-470. https://doi.org/10.1063/1.351876

[143] Bording, J.K., Li, B.Q., Shi, Y.F., Zuo, J.M., Physical Review Letters, 90[22] 2003, 226104. https://doi.org/10.1103/PhysRevLett.90.226104

[144] Naik, R., Kota, C., Rao, B.U.M., Journal of Vacuum Science and Technology A, 12[4] 1994, 1832-1837. https://doi.org/10.1116/1.579013

[145] Kato, M., Niwa, H., Philosophical Magazine B, 64[3] 1991, 317-326. https://doi.org/10.1080/13642819108207622

[146] Yapsir, A.S., Choi, C.H., Lu, T.M., Journal of Applied Physics, 67[2] 1990, 796-799. https://doi.org/10.1063/1.345734

[147] Liu, H., Zhang, Y.F., Wang, D.Y., Pan, M.H., Jia, J.F., Xue, Q.K., Surface Science, 571[1-3] 2004, 5-11. https://doi.org/10.1016/j.susc.2004.08.011

[148] Liu, H., Zhang, Y.F., Wang, D.Y., Jia, J.F., Xue, Q.K., Chinese Physics Letters, 21[8] 2004, 1608-1611.

[149] Hasan, M.A., Radnoczi, G., Sundgren, J.E., Hansson, G.V., Surface Science, 236[1-2] 1990, 53-76. https://doi.org/10.1016/0039-6028(90)90761-V

[150] Westmacott, K.H., Hinderberger, S., Dahmen, U., Philosophical Magazine A, 81[6] 2001, 1547-1578. https://doi.org/10.1080/01418610108214362

[151] Hsieh, Y.F., Chen, L.J., Marshall, E.D., Lau, S.S., Thin Solid Films, 162, 1988, 287-294. https://doi.org/10.1016/0040-6090(88)90217-9

[152] Roy, A., Bhattacharjee, K., Dev, B.N., Applied Surface Science, 256[2] 2009, 508-512. https://doi.org/10.1016/j.apsusc.2009.07.085

[153] Horn-von Hoegen, M., Henzler, M., Physica Status Solidi A, 146[1] 1994, 337-352. https://doi.org/10.1002/pssa.2211460129

[154] Ernst, F., Philosophical Magazine A, 68[6] 1993, 1251-1272. https://doi.org/10.1080/01418619308222930

[155] Ernst, F., Materials Research Society Symposium - Proceedings, 319, 1994, 165-170.

[156] Cao, S.P., Ye, F., Xu, A.Y., Bai, F.M., Materials Research Innovations, 18, 2014, S4642-S4645.

[157] Shinkai, S., Sasaki, K., Japanese Journal of Applied Physics - 1, 38[6A] 1999, 3646-3650. https://doi.org/10.1143/JJAP.38.3646

[158] Oishi, N., Yanagisawa, H., Sasaki, K., Abe, Y., Kawamura, M., Electronics and Communications in Japan - II, 81[9] 1998, 46-52.
https://doi.org/10.1002/(SICI)1520-6432(199809)81:9<46::AID-ECJB6>3.0.CO;2-Z

[159] Kaushik, V.S., Datye, A.K., Kendall, D.L., Martinez-Tovar, B., Myers, D.R., Applied Physics Letters, 52[21] 1988, 1782-1784. https://doi.org/10.1063/1.99722

[160] Kutana, A., Erwin, S.C., Physical Review B, 87[4] 2013, 045314.
https://doi.org/10.1103/PhysRevB.87.045314

[161] Jnawali, G., Hattab, H., Meyer Zu Heringdorf, F.J., Krenzer, B., Horn-Von Hoegen, M., Physical Review B, 76[3] 2007, 035337.
https://doi.org/10.1103/PhysRevB.76.035337

[162] Furdyna, J.K., Kossut, J., Superlattices and Microstructures, 2[1] 1986, 89-96.
https://doi.org/10.1016/0749-6036(86)90160-6

[163] Cunningham, J.E., Pathak, R.N., Jan, W.Y., Applied Physics Letters, 68[3] 1996, 394-396. https://doi.org/10.1063/1.116696

[164] Jones, K.A., Tu, C.W., Journal of Crystal Growth, 70[1-2] 1984, 127-132.
https://doi.org/10.1016/0022-0248(84)90257-4

[165] Allovon, M., Primot, J., Gao, Y., Quillec, M., Journal of Electronic Materials, 18[4] 1989, 505-510. https://doi.org/10.1007/BF02657780

[166] Sagalowicz, L., Jouneau, P.H., Rudra, A., Syrbu, A.V., Kapon, E., Proceedings of the TMS Fall Meeting, 1998, 259-268.

[167] Xu, B., He, P., Liu, H., Wang, P., Zhou, G., Wang, X., Angewandte Chemie, 53[9] 2014, 2339-2343. https://doi.org/10.1002/anie.201310513

[168] Kozlovsky, V.I., Martovitsky, V.P., Skasyrsky, Y.K., Sadofyev, Y.G., Turyansky, A.G., Physica Status Solidi B, 229[1] 2002, 63-67. https://doi.org/10.1002/1521-3951(200201)229:1<63::AID-PSSB63>3.0.CO;2-8

[169] Tobin, S.P., Smith, F.T.J., Norton, P.W., Wu, J., Dudley, M., Marzio, D.D.,

Casagrande, L.G., Journal of Electronic Materials, 24[9] 1995, 1189-1199. https://doi.org/10.1007/BF02653073

[170] Maekawa, T., Saito, T., Yoshikawa, M., Takigawa, H., Materials Research Society Symposia Proceedings, 56, 1986, 109-113. https://doi.org/10.1557/PROC-56-109

[171] Uemoto, T., Kamata, A., Mitsuhashi, H., Hirahara, K., Beppu, T., Journal of Crystal Growth, 99[1-4] 1990, 422-426. https://doi.org/10.1016/0022-0248(90)90556-Z

[172] Fujita, S., Terada, K., Sakamoto, T., Fujita, S., Journal of Crystal Growth, 94[1] 1989, 102-108. https://doi.org/10.1016/0022-0248(89)90608-8

[173] Ohmi, K., Suemune, I., Kanda, T., Kan, Y., Yamanishi, M., Japanese Journal of Applied Physics, 26[12A] 1987, L2072-L2075. https://doi.org/10.1143/JJAP.26.L2072

[174] Matsumura, N., Ishikawa, K., Saraie, J., Yodogawa, Y., Journal of Crystal Growth, 72[1-2] 1985, 41-45. https://doi.org/10.1016/0022-0248(85)90115-0

[175] Funato, M., Kitani, H., Fujita, S., Fujita, S., Journal of Electronic Materials, 25[2] 1996, 217-222. https://doi.org/10.1007/BF02666247

[176] Kečkéš, J., Ortner, B., Červeň, I., Jakabovič, J., Kováč, J., Journal of Applied Physics, 80[11] 1996, 6204-6210. https://doi.org/10.1063/1.363696

[177] Sadowski, J., Herman, M.A., Journal of Crystal Growth, 146[1-4] 1995, 449-454. https://doi.org/10.1016/0022-0248(94)00512-5

[178] Rotter, S., Kasemset, D., Fonstad, C.G., IEEE Electron Device Letters, 3[3] 1982, 66-68. https://doi.org/10.1109/EDL.1982.25481

[179] Shigenaka, K., Sugiura, L., Nakata, F., Hirahara, K., Journal of Crystal Growth, 145[1-4] 1994, 376-381. https://doi.org/10.1016/0022-0248(94)91079-0

[180] Ivanov, I.S., Sidorov, Y.G., Yakushev, M.V., Inorganic Materials, 33[3] 1997, 243-247. https://doi.org/10.1017/S0034412597003879

[181] Sakurai, T., Matsumoto, N., Okada, Y., Onari, S., Akimoto, K., Physica Status Solidi C, 2[7] 2005, 2224-2227. https://doi.org/10.1002/pssc.200461460

[182] Beanland, R., Kiely, C.J., Interface Science, 1[2] 1993, 99-113. https://doi.org/10.1007/BF00203599

[183] Hung, L.S., Zheng, L.R., Blanton, T.N., Applied Physics Letters, 60[25] 1992, 3129-3131. https://doi.org/10.1063/1.106745

[184] Rajan, K., Gong, R., Webb, J., Applied Physics Letters, 57[14] 1990, 1446-1448. https://doi.org/10.1063/1.103365

[185] Ortner, B., Bauer, G., Journal of Crystal Growth, 92[1-2] 1988, 69-76. https://doi.org/10.1016/0022-0248(88)90435-6

[186] Shiau, F.Y., Chang, Y.A., Chen, L.J., Journal of Electronic Materials, 17[5] 1988, 433-441. https://doi.org/10.1007/BF02652130

[187] Kenty, J.L., Thin Solid Films, 26[1] 1975, 181-195. https://doi.org/10.1016/0040-6090(75)90177-7

[188] Golding, T.D., Dura, J.A., Wang, W.C., Zborowski, J.T., Vigliante, A., Chen, H.C., Meyer, J.R., Journal of Crystal Growth, 127[1-4] 1993, 777-782. https://doi.org/10.1016/0022-0248(93)90731-B

[189] Golding, T.D., Dura, J.A., Wang, H., Zborowski, J.T., Vigliante, A., Chen, H.C., Miller, J.H., Meyer, J.R., Semiconductor Science and Technology, 8[1S] 1993, S117-S120. https://doi.org/10.1088/0268-1242/8/1S/026

[190] Golding, T.D., Dura, J.A., Wang, W.C., Vigliante, A., Moss, S.C., Chen, H.C., Miller, J.H., Hoffman, C.A., Meyer, J.R., Applied Physics Letters, 63[8] 1993, 1098-1100. https://doi.org/10.1063/1.109792

[191] Hsieh, J.J., Finn, M.C., Rossi, J.A., Institute of Physics - Conference Series, 33, 1976, 37-44.

[192] Wright, P.D., Rezek, E.A., Holonyak, N., Stillman, G.E., Rossi, J.A., Groves, W.O., Applied Physics Letters, 31[1] 1977, 40-42. https://doi.org/10.1063/1.89472

[193] Wright, P.D., Rezek, E.A., Holonyak, N., Journal of Crystal Growth, 41[2] 1977, 254-261. https://doi.org/10.1016/0022-0248(77)90053-7

[194] Tamura, A., Oka, K., Inoue, M., Shirafuji, J., Inuishi, Y., Proceedings of the Conference on Solid State Devices, 1980, 479-448. https://doi.org/10.7567/JJAPS.19S1.479

[195] Tamura, A., Oka, K., Inoue, M., Shirafuji, J., Inuishi, Y., Japanese Journal of Applied Physics, 19, 1980, 479-482. https://doi.org/10.7567/JJAPS.19S1.479

[196] Efimov, A.N., Lebedev, A.O., Tsaregorodtsev, A.M., Journal of Applied Crystallography, 31[3] 1998, 461-473. https://doi.org/10.1107/S0021889897011801

[197] Zhylik, A., Benediktovitch, A., Feranchuk, I., Inaba, K., Mikhalychev, A., Ulyanenkov, A., Journal of Applied Crystallography, 46[4] 2013, 919-925.

https://doi.org/10.1107/S0021889813006171

[198] Kato, K., Sasaki, K., Abe, Y., Japanese Journal of Applied Physics - 1, 45[4A] 2006, 2731-2735. https://doi.org/10.1143/JJAP.45.2731

[199] Kato, K., Sasaki, K., Abe, Y., Japanese Journal of Applied Physics - 1, 45[9A] 2006, 7097-7099. https://doi.org/10.1143/JJAP.45.7097

[200] Seong, W.K., Oh, S., Kang, W.N., Japanese Journal of Applied Physics, 51[8-1] 2012, 083101. https://doi.org/10.7567/JJAP.51.083101

[201] Koyama, T., Chichibu, S.F., Journal of Applied Physics, 95[12] 2004, 7856-7861. https://doi.org/10.1063/1.1739294

[202] Morita, K., Tsurekawa, S., Nakashima, H., Yoshinaga, H., Journal of the Japan Institute of Metals, 59[9] 1995, 881-888. https://doi.org/10.2320/jinstmet1952.59.9_881

[203] Stock, S.R., Ahn, S.H., Cohen, J.B., Journal of the American Ceramic Society, 70[3] 1987, 125-132. https://doi.org/10.1111/j.1151-2916.1987.tb04946.x

[204] Ding, H., Dwaraknath, S.S., Garten, L., Ndione, P., Ginley, D., Persson, K.A., ACS Applied Materials and Interfaces, 8, 2016, 13086–13093. https://doi.org/10.1021/acsami.6b01630

[205] Watanabe, T., Saito, K., Osada, M., Suzuki, T., Fujimoto, M., Yoshimoto, M., Sasaki, A., Liu, J., Kakihana, M., Funakubo, H., Materials Research Society Symposium - Proceedings, 748, 2003, 69-74. https://doi.org/10.1557/PROC-748-U2.4

[206] Watanabe, T., Funakubo, H., Saito, K., Suzuki, T., Fujimoto, M., Osada, M., Noguchi, Y., Miyayama, M., Applied Physics Letters, 81[9] 2002, 1660-1662. https://doi.org/10.1063/1.1503850

[207] Vuorinen, S., Hoel, R.H., Thin Solid Films, 232[1] 1993, 73-82. https://doi.org/10.1016/0040-6090(93)90765-H

[208] Tang, Y., Dai, F., Gu, X., Wang, Z., Zhang, W., Physica E, 77, 2016, 97-101. https://doi.org/10.1016/j.physe.2015.11.007

[209] Jia, C.L., Hojczyk, R., Faley, M., Poppe, U., Urban, K., Philosophical Magazine A, 79[4] 1999, 873-891. https://doi.org/10.1080/01418619908210337

[210] Chiba, K., Makino, S., Mukaida, M., Kusunoki, M., Ohshima, S., IEEE Transactions on Applied Superconductivity, 11[1] 2001, 2734-2737. https://doi.org/10.1109/77.919628

[211] Ito, W., Yoshida, Y., Mahajan, S., Morishita, T., Journal of Crystal Growth, 146[1-4] 1995, 655-658. https://doi.org/10.1016/0022-0248(94)00500-1

[212] Vignolle, C., Gervais, A., Physica Status Solidi A, 126[1] 1991, 197-203. https://doi.org/10.1002/pssa.2211260122

[213] Lee, S.T., Chen, S., Hung, L.S., Braunstein, G., Applied Physics Letters, 55[3] 1989, 286-288. https://doi.org/10.1063/1.102406

[214] Shi, L., Zhou, G., Jia, Y., Huang, Y., Wang, R.L., Wang, C.A., Yi, H.A., Li, H.C., Zhang, Y., Superconductor Science and Technology, 6[3] 1993, 191-194. https://doi.org/10.1088/0953-2048/6/3/005

[215] Inoue, K., Saito, M., Wang, Z., Kotani, M., Ikuhara, Y., Materials Transactions, 56[3] 2015, 281-287. https://doi.org/10.2320/matertrans.M2014394

[216] Hay, R.S., Acta Materialia, 55[3] 2007, 991-1007. https://doi.org/10.1016/j.actamat.2006.09.029

[217] Ferdeghini, C., Grassano, G., Bellingeri, E., Marrè, D., Ramadan, W., Ferrando, V., Beneduce, C., International Journal of Modern Physics B, 17[4-6 II] 2003, 824-829. https://doi.org/10.1142/S0217979203016674

[218] Gorbenko, O.Y., Bosak, A.A., Journal of Crystal Growth, 186[1-2] 1998, 181-188. https://doi.org/10.1016/S0022-0248(97)00454-5

[219] Turan, S., Knowles, K.M., Interface Science, 8[2] 2000, 279-294. https://doi.org/10.1023/A:1008724002737

[220] Peng, Y., Han, G., Wang, D., Wang, K., Guo, Z., Yang, J., Yuan, W., International Journal of Hydrogen Energy, 42[21] 2017, 14409-14417. https://doi.org/10.1016/j.ijhydene.2017.04.204

[221] Xu, Z., Salvador, P., Kitchin, J.R., ACS Applied Materials and Interfaces, 9[4] 2017, 4106-4118. https://doi.org/10.1021/acsami.6b11791

[222] Ohno, T., Ii, S., Shibata, N., Matsunaga, K., Ikuhara, Y., Yamamoto, T., Journal of the Japan Institute of Metals, 69[11] 2005, 1004-1009. https://doi.org/10.2320/jinstmet.69.1004

[223] Ohno, T., Ii, S., Shibata, N., Matsunaga, K., Ikuhara, Y., Yamamoto, T., Materials Transactions, 45[7] 2004, 2117-2121. https://doi.org/10.2320/matertrans.45.2117

[224] Bristow, J.K., Butler, K.T., Svane, K.L., Gale, J.D., Walsh, A., Journal of Materials Chemistry A, 5[13] 2017, 6226-6232. https://doi.org/10.1039/C7TA00356K

[225] Ding, H., Dwaraknath, S.S., Garten, L., Ndione, P., Ginley, D., Persson, K.A., ACS

Applied Materials and Interfaces, 8[20] 2016, 13086-13093.
https://doi.org/10.1021/acsami.6b01630

[226] Liu, H., Cheng, X., Valanoor, N., ACS Applied Materials and Interfaces, 8[50] 2016, 34844-34853. https://doi.org/10.1021/acsami.6b10701

[227] Wu, H., Chen, O., Zhuang, J., Lynch, J., Lamontagne, D., Nagaoka, Y., Cao, Y.C., Journal of the American Chemical Society, 133[36] 2011, 14327-14337. https://doi.org/10.1021/ja2023724

[228] Sakellari, D., Frangis, N., Polychroniadis, E.K., Physica E, 42[5] 2010, 1777-1780. https://doi.org/10.1016/j.physe.2010.01.047

[229] Lau, Y.K.A., Chernak, D.J., Bierman, M.J., Jin, S., Journal of Materials Chemistry, 19[7] 2009, 934-940. https://doi.org/10.1039/b818187j

[230] Mohanty, D., Sun, X., Lu, Z., Washington, M., Wang, G.C., Lu, T.M., Bhat, I.B., Journal of Applied Physics, 124[17] 2018, 175301. https://doi.org/10.1063/1.5052644

[231] Littlejohn, A.J., Xiang, Y., Rauch, E., Lu, T.M., Wang, G.C., Journal of Applied Physics, 122[18] 2017, 185305. https://doi.org/10.1063/1.5000502

[232] Gwo, S., Wu, C.L., Shen, C.H., Lin, H.W., Chen, H.Y., Ahn, H., Proceedings of SPIE - The International Society for Optical Engineering, 6134, 2006, 61340L.

[233] Wu, C.L., Shen, C.H., Lin, H.W., Lee, H.M., Gwo, S., Applied Physics Letters, 87[24] 2005, 241916. https://doi.org/10.1063/1.2146062

[234] Kamohara, T., Akiyama, M., Ueno, N., Kuwano, N., Ceramics International, 34[4] 2008, 985-989. https://doi.org/10.1016/j.ceramint.2007.09.051

[235] Kamohara, T., Akiyama, M., Ueno, N., Nonaka, K., Kuwano, N., Applied Physics Letters, 89[7] 2006, 071919. https://doi.org/10.1063/1.2337558

[236] Shinkai, S., Sasaki, K., Yanagisawa, H., Yoshio, A., Japanese Journal of Applied Physics - 1, 42[10] 2003, 6518-6522. https://doi.org/10.1143/JJAP.42.6518

[237] Abe, Y., Kawamura, M., Sasaki, K., Japanese Journal of Applied Physics - 1, 41[11B] 2002, 6857-6861. https://doi.org/10.1143/JJAP.41.6857

[238] Efimov, A.N., Lebedev, A.O., Crystallography Reports, 47[1] 2002, 135-144. https://doi.org/10.1134/1.1446923

[239] Lee, J.J., Kang, K.Y., Park, Y.S., Yang, C.S., Kim, H.S., Klm, K.H., Kang, T.W., Park, S.H., Lee, J.Y., Japanese Journal of Applied Physics - 1, 38[11] 1999, 6487-6488. https://doi.org/10.1143/JJAP.38.6487

[240] Lee, J.J., Park, Y.S., Yang, C.S., Kim, H.S., Kim, K.H., Kang, K.Y., Kang, T.W., Park, S.H., Lee, J.Y., Journal of Crystal Growth, 213[1] 2000, 33-39. https://doi.org/10.1016/S0022-0248(00)00335-3

[241] Ito, A., Masumoto, H., Goto, T., Thin Solid Films, 517[19] 2009, 5616-5620. https://doi.org/10.1016/j.tsf.2009.02.021

[242] Sandström, P., Svedberg, E.B., Birch, J., Sundgren, J.E., Journal of Crystal Growth, 197[4] 1999, 849-857. https://doi.org/10.1016/S0022-0248(98)00972-5

[243] Svedberg, E.B., Sandström, P., Sundgren, J.E., Greene, J.E., Madsen, L.D., Surface Science, 429[1] 1999, 206-216. https://doi.org/10.1016/S0039-6028(99)00379-9

[244] McCaffrey, J.P., Svedberg, E.B., Phillips, J.R., Madsen, L.D., Journal of Crystal Growth, 200[3] 1999, 498-504. https://doi.org/10.1016/S0022-0248(98)01403-1

[245] Shoup, S.S., Paranthaman, M., Goyal, A., Specht, E.D., Lee, D.F., Kroeger, D.M., Beach, D.B., Journal of the American Ceramic Society, 81[11] 1998, 3019-3021. https://doi.org/10.1111/j.1151-2916.1998.tb02731.x

[246] Unal, O., Mitchell, T.E., Journal of Materials Research, 7[6] 1992, 1445-1454. https://doi.org/10.1557/JMR.1992.1445

[247] Ren, S.Y., Dow, J.D., Applied Physics Letters, 69[2] 1996, 251-253. https://doi.org/10.1063/1.117940

[248] Wang, S., Xu, X., Luo, H., Cao, C., Song, X., Zhao, J., Zhang, J., Tang, C., RSC Advances, 8[34] 2018, 19279-19288. https://doi.org/10.1039/C8RA02121J

[249] Foronda, H.M., Mazumder, B., Young, E.C., Laurent, M.A., Li, Y., DenBaars, S.P., Speck, J.S., Journal of Crystal Growth, 475, 2017, 127-135. https://doi.org/10.1016/j.jcrysgro.2017.06.008

[250] González, J.A., Andrés, J.P., López Antón, R., De Toro, J.A., Normile, P.S., Muniz, P., Riveiro, J.M., Nogués, J., Chemistry of Materials, 29[12] 2017, 5200-5206. https://doi.org/10.1021/acs.chemmater.7b00868

[251] Luo, S., Wang, C., Zhang, S., Tu, R., Liu, S., Tang, X., Shen, Q., Chen, F., Zhang, L., Applied Physics Express, 5[8] 2012, 085801. https://doi.org/10.1143/APEX.5.085801

[252] Sakai, Y., Saito, S., Cohen, M.L., Physical Review B, 89[11] 2014, 115424. https://doi.org/10.1103/PhysRevB.89.115424

[253] Li, Y., Xiong, H., Chen, G., Yan, Z., Ji, X., Xu, K., Miao, L., Proceedings - International Conference on Natural Computation, 2013, 6818090, 829-833.

[254] Yang, C., Chen, Z., Hu, J., Ren, Z., Lin, S., Materials Research Bulletin, 47[6] 2012, 1331-1334. https://doi.org/10.1016/j.materresbull.2012.03.018

[255] Li, L.B., Chen, Z.M., Xie, L.F., Yang, C., Journal of Crystal Growth, 385, 2014, 111-114. https://doi.org/10.1016/j.jcrysgro.2013.07.019

[256] Yang, C., Chen, Z., Li, L., Li, W., Hu, J., Lin, S., Solid State Communications, 152[2] 2012, 68-70. https://doi.org/10.1016/j.ssc.2011.10.037

[257] Aravazhi, S., Geskus, D., Van Dalfsen, K., Vázquez-Córdova, S.A., Grivas, C., Griebner, U., García-Blanco, S.M., Pollnau, M., Applied Physics B, 111[3] 2013, 433-446. https://doi.org/10.1007/s00340-013-5353-1

[258] Yin, Z., Zhang, P., Zhang, M.S., Applied Physics Letters, 68[16] 1996, 2303-2305. https://doi.org/10.1063/1.116171

[259] Bugakov, A.V., Ievlev, V.M., Physics, Chemistry and Mechanics of Surfaces, 10[12] 1995, 1457-1470.

[260] Homma, H., Yang, K.Y., Schuller, I.K., Physical Review B, 36[18] 1987, 9435-9438. https://doi.org/10.1103/PhysRevB.36.9435

[261] Schuller, I.K., Superlattices and Microstructures, 4[4-5] 1988, 521-524. https://doi.org/10.1016/0749-6036(88)90230-3

[262] Kaneko, T., Imafuku, M., Yamamoto, R., Doyama, M., Transactions of the Japan Institute of Metals, Supplement, 27, 1986, 323-328.

[263] Sakata, T., Yasuda, H.Y., Umakoshi, Y., Acta Materialia, 51[6] 2003, 1561-1572. https://doi.org/10.1016/S1359-6454(02)00558-X

[264] Xu, W.S., Zhang, W.Z., Philosophical Magazine, 98[1] 2018, 75-93. https://doi.org/10.1080/14786435.2017.1390619

[265] Guziewski, M., Coleman, S.P., Weinberger, C.R., Acta Materialia, 155, 2018, 1-11. https://doi.org/10.1016/j.actamat.2018.05.051

[266] Guziewski, M., Coleman, S.P., Weinberger, C.R., Acta Materialia, 119, 2016, 184-192. https://doi.org/10.1016/j.actamat.2016.08.017

[267] Ye, F., Zhang, W.Z., Qiu, D., Acta Materialia, 54[20] 2006, 5377-5384. https://doi.org/10.1016/j.actamat.2006.07.006

[268] Han, S.Z., Park, S.I., Huh, J.S., Lee, Z.H., Lee, H.M., Materials Science and Engineering A, 230[1-2] 1997, 100-106. https://doi.org/10.1016/S0921-5093(96)10857-1

Materials Research Forum LLC
https://doi.org/10.21741/9781644900475

[269] Zhou, J.P., Zhao, D.S., Zheng, O., Wang, J.B., Xiong, D.X., Sun, Z.F., Gui, J.N., Wang, R.H., Micron, 40[8] 2009, 906-910. https://doi.org/10.1016/j.micron.2009.05.008

[270] Li, Y.J., Zhang, W.Z., Marthinsen, K., Acta Materialia, 60[17] 2012, 5963-5974. https://doi.org/10.1016/j.actamat.2012.06.022

[271] Matsuda, Y., Sakamoto, K., Yahisa, Y., Hosoe, Y., Hosoda, H., Kitamoto, Y., Journal of Magnetism and Magnetic Materials, 469, 2019, 545-549. https://doi.org/10.1016/j.jmmm.2018.09.001

[272] Douin, J., Dahmen, U., Westmacott, K.H., Philosophical Magazine B, 63[4] 1991, 867-890. https://doi.org/10.1080/13642819108205543

[273] Budai, J.D., Young, R.T., Chao, B.S., Applied Physics Letters, 62[15] 1993, 1836-1838. https://doi.org/10.1063/1.109565

[274] Knorr, D.B., Merchant, S.M., Biberger, M.A., Journal of Vacuum Science and Technology B, 16[5] 1998, 2734-2744. https://doi.org/10.1116/1.590265

[275] Fartash, A., Thin Solid Films, 323[1-2] 1998, 296-303. https://doi.org/10.1016/S0040-6090(97)00931-0

[276] Je, J.H., You, H., Cullen, W.G., Maroni, V.A., Ma, B., Koritala, R.E., Rupich, M.W., Thieme, C.L.H., Physica C, 384[1-2] 2003, 54-60. https://doi.org/10.1016/S0921-4534(02)01982-2

[277] Pantleon, K., Somers, M.A.J., Acta Materialia, 52[16] 2004, 4929-4940. https://doi.org/10.1016/j.actamat.2004.07.001

[278] Pantleon, K., Somers, M.A.J., Materials Science Forum, 495-497[2] 2005, 1455-1460. https://doi.org/10.4028/www.scientific.net/MSF.495-497.1455

[279] Lu, Z., Sun, X., Washington, M.A., Lu, T.M., Journal of Physics D, 51[9] 2018, 095301. https://doi.org/10.1088/1361-6463/aaa875

[280] Chen, J.K., Reynolds, W.T., Acta Materialia, 45[11] 1997, 4423-4430. https://doi.org/10.1016/S1359-6454(97)00158-4

[281] Tsurekawa, S., Tanaka, T., Nakashima, H., Yoshinaga, H., Journal of the Japan Institute of Metals, 58[4] 1994, 377-381. https://doi.org/10.2320/jinstmet1952.58.4_377

[282] Sasajima, Y., Yamamoto, R., Doyama, M., Transactions of the Japan Institute of Metals, Supplement, 27, 1986, 301-306.

[283] Seki, A., Kame, K., Journal of the Iron and Steel Institute of Japan, 77[7] 1991,

892-897. https://doi.org/10.2355/tetsutohagane1955.77.7_892

[284] Seki, A., Kame, K., ISIJ International, 32[12] 1992, 1306-1310. https://doi.org/10.2355/isijinternational.32.1306

[285] Chen, J.K., Ross, T.W., Chen, G., Kikuchi, M., Reynolds, W.T., Metallurgical and Materials Transactions A, 25[12] 1994, 2639-2646. https://doi.org/10.1007/BF02649217

[286] Yang, J., Yang, Z., Qiu, D., Zhang, W., Zhang, C., Bai, B., Fang, H., Acta Metallurgica Sinica, 41[3] 2005, 225-230.

[287] Zhang, W.Z., Weatherly, G.C., Acta Materialia, 46[6] 1998, 1837-1847. https://doi.org/10.1016/S1359-6454(97)00435-7

[288] Zhang, W.Z., Wu, J., Materials Science and Engineering A, 438-440[S] 2006, 118-121. https://doi.org/10.1016/j.msea.2006.01.100

[289] Morita, K., Uehara, M., Tsurekawa, S., Nakashima, H., Journal of the Japan Institute of Metals, 61[4] 1997, 251-260. https://doi.org/10.2320/jinstmet1952.61.4_251

[290] Wang, W., Cai, C., Rohrer, G.S., Gu, X., Lin, Y., Chen, S., Dai, P., Materials Characterization, 144, 2018, 411-423. https://doi.org/10.1016/j.matchar.2018.07.040

[291] Shao, X.H., Jin, Q.Q., Zhou, Y.T., Yang, H.J., Zheng, S.J., Zhang, B., Chen, Q., Ma, X.L., Materialia, 6, 2019, 100287. https://doi.org/10.1016/j.mtla.2019.100287

[292] Lanxner, M., Bauer, C.L., Transactions of the Japan Institute of Metals, Supplement, 27, 1986, 617-624.

[293] Onda, T., Piao, M., Bando, Y., Ichinose, H., Otsuka, K., Materials Transactions, JIM, 36[1] 1995, 23-29. https://doi.org/10.2320/matertrans1989.36.23

[294] Wang, Y.G., Zhang, Z., Yan, G.H., De Hosson, J.T.M., Journal of Materials Science, 37[12] 2002, 2511-2518. https://doi.org/10.1023/A:1015439624193

[295] Meissner, M., Sojka, F., Matthes, L., Bechstedt, F., Feng, X., Müllen, K., Mannsfeld, S.C.B., Forker, R., Fritz, T., ACS Nano, 10[7] 2016, 6474-6483. https://doi.org/10.1021/acsnano.6b00935

[296] Ward, M.D., ACS Nano, 10[7] 2016, 6424-6428. https://doi.org/10.1021/acsnano.6b03830

[297] Simbrunner, C., Schwabegger, G., Resel, R., Dingemans, T., Quochi, F., Saba, M., Mura, A., Bongiovanni, G., Sitter, H., Crystal Growth and Design, 14[11] 2014,

5719-5728. https://doi.org/10.1021/cg500979p

[298] Sun, X., Lu, Z., Xie, W., Wang, Y., Shi, J., Zhang, S., Washington, M.A., Lu, T.M., Applied Physics Letters, 110[15] 2017, 153104. https://doi.org/10.1063/1.4980088

[299] Littlejohn, A.J., Lu, T.M., Zhang, L.H., Kisslinger, K., Wang, G.C., CrystEngComm, 18[15] 2016, 2757-2769. https://doi.org/10.1039/C5CE02579F

[300] Liang, J.J., Kung, P.W.C., Journal of Materials Research, 17[7] 2002, 1686-1691. https://doi.org/10.1557/JMR.2002.0248

[301] Wallace, S.K., Butler, K.T., Hinuma, Y., Walsh, A., Journal of Applied Physics, 125[5] 2019, 055703. https://doi.org/10.1063/1.5079485

[302] Beyer, P., Breuer, T., Ndiaye, S., Zykov, A., Viertel, A., Gensler, M., Rabe, J.P., Hecht, S., Witte, G., Kowarik, S., ACS Applied Materials and Interfaces, 6[23] 2014, 21484-21493. https://doi.org/10.1021/am506465b

Keywords

0-lattice, 5, 16-21, 71, 77, 90-91, 107-108, 115-117

anisotropy, 23, 58, 86, 109, 120
annealing twins, 8
areal ratio, 7, 29
atomic bonding, 5, 19, 109
austenite, 107, 116
axial ratio, 6-7
axis of rotation, 2
azimuthal, 45, 50, 62, 119

Bagaryatskii, 107-108
ball-packing, 3
best match, 15-16, 43, 80, 105
bicrystal, 10-11, 14, 87
body-centered tetragonal, 7

Bollmann, 15, 125-127
Bravais, 89
buffer, 3, 25, 68, 79, 85, 90
Burgers, 5, 7-11, 14-15, 17, 19-21, 42, 49, 50-51, 54-55, 59, 72, 80-81, 90, 106, 108, 115, 117

carrier mobility, 2
cementite, 107, 108
chemical energy, 23
coding theory, 3
coincidence site lattice, 2, 5-7, 9, 11-12, 14-16, 22-23, 25, 32-33, 65, 81, 84, 87, 90, 92, 106-108, 115
coinciding lattice sites, 2
crosshatch, 78
crystal structure, 2-3, 14-17, 26, 32-34, 38, 42, 86, 89
crystal symmetry, 3, 44
Curie law, 33, 84

Diophantine problem, 4
displacement shift complete, 5, 7-15, 17, 45, 87, 93, 106, 115
domain structure, 3, 52

electron energies, 3
electronegativity, 30, 31
energy-band alignment, 3
epitaxial strain, 3
epitaxial thin film, 101
Eshelby, 33
Euclidean distance, 4
eutectic, 14, 23, 68, 92, 117

ferrite, 107-108
fit-misfit, 5
Frank-Bilby, 20

generating function, 7
geometry of numbers, 3-4
grain-boundary dislocation, 7, 9, 11, 14, 19
grain-boundary structure, 4-7, 9, 12, 19-20, 87, 93, 117

habit plane, 14, 17, 20, 90, 108-109, 116
Heine-Abarenkov, 115
heterostructural, 3, 13-14, 18, 21, 26, 32, 80, 90, 94
hexagonal lattice, 6, 17, 25

impurity effects, 2
incommensurate interfaces, 35
integral lattice point, 4
intentionally strained, 2
interface energy, 2, 64
interfacial coupling, 3
interfacial properties, 3
interpenetrating lattices, 15-16
interphase boundaries, 5, 9, 17, 92, 106-107, 115
interplanar spacing, 23, 46, 80
ionic conductivity, 3, 121
Isaichev, 107

kissing-number, 3
Kurdjumov-Sachs, 20, 107, 115

lattice constants, 3, 32, 36, 72, 77, 85-87, 91, 95, 98, 106, 108, 122

About the author

Dr Fisher has wide knowledge and experience of the fields of engineering, metallurgy and solid-state physics, beginning with work at Rolls-Royce Aero Engines on turbine-blade research, related to the Concord supersonic passenger-aircraft project, which led to a BSc degree (1971) from the University of Wales. This was followed by theoretical and experimental work on the directional solidification of eutectic alloys having the ultimate aim of developing composite turbine blades. This work led to a doctoral degree (1978) from the Swiss Federal Institute of Technology (Lausanne). He then acted for many years as an editor of various academic journals, in particular *Defect and Diffusion Forum*. In recent years he has specialised in writing monographs which introduce readers to the most rapidly developing ideas in the fields of engineering, metallurgy and solid-state physics. His latest paper will appear shortly in *International Materials Reviews*, and he is co-author of the widely-cited student textbook, *Fundamentals of Solidification*.

[1] Rosciano, F., Pescarmona, P.P., Houthoofd, K., Persoons, A., Bottke, P., Wilkening, M., Physical Chemistry Chemical Physics, 15[16] 2013, 6107-6112.

[2] Zou, M., Yang, F., Wen, K., Lv, W., Waqas, M., He, W., International Journal of Hydrogen Energy, 41[47] 2016, 22254-22259.

[3] Liao, Z., Green, R.J., Gauquelin, N., Macke, S., Li, L., Gonnissen, J., Sutarto, R., Houwman, E.P., Zhong, Z., Van Aert, S., Verbeeck, J., Sawatzky, G.A., Huijben, M., Koster, G., Rijnders, G., Advanced Functional Materials, 26[36] 2016, 6627-6634.

[4] Butler, K.T., Hendon, C.H., Walsh, A., Faraday Discussions, 201, 2017, 207-219.

[5] Davenport, H., The Mathematical Gazette, 31[296] 1947, 206-210.

[6] Friedel, G., Lecons de Cristallographie, Blenchard, 1926.

[7] Balluffi, R.W., Brokman, A., King, A.H., Acta Metallurgica, 30[8] 1982, 1453-1470.

[8] Morita, K., Tsurekawa, S., Nakashima, H., Yoshinaga, H., Materials Science Forum, 204-206[1] 1996, 239-244

[9] Miyano, N., Ameyama, K., Journal of the Japan Institute of Metals, 64[1] 2000, 42-49.

[10] King, A.H., Singh, A., Journal of Physics and Chemistry of Solids, 55[10] 1994, 1023-1033.

[11] Shiue, Y.R., Phillips, D.S., Phillips, D.S., Philosophical Magazine A, 50[5] 1984, 677-702.

[12] Fukutomi, Hiroshi, Tanaka, Mutsuto, Nippon Kinzoku Gakkaisi, 49[8] 1985, 607-613.

[13] Zhu, Y., Suenaga, M., Philosophical Magazine A, 66[3] 1992, 457-471.

[14] Chen, L.Q., Kalonji, G., Philosophical Magazine A, 66[1] 1992, 11-26.

[15] Thibault, J., Putaux, J.L., Jacques, A., George, A., Michaud, H.M., Baillin, X., Materials Science and Engineering A, 164[1-2] 1993, 93-100.

[16] Mou, Y., Metallurgical and Materials Transactions A, 25[9] 1994, 1905-1915.

[17] Morniroli, J.P., Cherns, D., Ultramicroscopy, 62[1-2] 1996, 53-63.

[18] Antoniades, I.P., Bleris, G.L., Philosophical Magazine A, 80[12] 2000, 2871-2897.

[19] Poulat, S., Thibault, J., Priester, L., Interface Science, 8[1] 2000, 5-15.

[20] Gertsman, V.Y., Acta Crystallographica A, 58[2] 2002, 155-161.

[21] Hu, J.R., Chang, S.C., Chen, F.R., Kai, J.J., Materials Chemistry and Physics, 74[3] 2002, 313-319.

[22] Couzinié, J.P., Decamps, B., Priester, L., Philosophical Magazine Letters, 83[12] 2003, 721-731.

[23] Ikuhara, Y., Nishimura, H., Nakamura, A., Matsunaga, K., Yamamoto, T., Peter, K., Lagerlöf, D., Journal of the American Ceramic Society, 86[4] 2003, 595-602.

[24] Décamps, B., Priester, L., Thibault, J., Advanced Engineering Materials, 6[10] 2004, 814-818.

[25] Hyde, B., Farkas, D., Caturla, M.J., Philosophical Magazine, 85[32] 2005, 3795-3807.

[26] Kizuka, T., Japanese Journal of Applied Physics - 1, 46[11] 2007, 7396-7398.

[27] Mompiou, F., Legros, M., Caillard, D., Materials Research Society Symposium Proceedings, 1086, 2008, 19-24.

[28] Sheikh-Ali, A.D., Acta Materialia, 58[19] 2010, 6249-6255.

[29] Mitsuma, T., Tohei, T., Shibata, N., Mizoguchi, T., Yamamoto, T., Ikuhara, Y., Journal of Materials Science, 46[12] 2011, 4162-4168.

[30] Tsuru, T., Kaji, Y., Shibutani, Y., Journal of Applied Physics, 110[7] 2011, 073520.

[31] Wan, L., Li, J., Modelling and Simulation in Materials Science and Engineering, 21[5] 2013, 055013.

[32] Saito, M., Wang, Z., Tsukimoto, S., Ikuhara, Y., Journal of Materials Science, 48[16] 2013, 5470-5474.

[33] Wan, L., Han, W., Chen, K., Scientific Reports, 5, 2015, 13441.

[34] Ishikawa, R., Lugg, N.R., Inoue, K., Sawada, H., Taniguchi, T., Shibata, N., Ikuhara, Y., Scientific Reports, 6, 2016, 21273.

[35] Sharma, N.K., Shekhar, S., Philosophical Magazine, 97[23] 2017, 2004-2017.

[36] Annevelink, E., Ertekin, E., Johnson, H.T., Acta Materialia, 166, 2019, 67-74.

[37] Potin, V., Ruterana, P., Nouet, G., Pond, R., Physical Review B, 61[8] 2000, 5587-5599.

[38] Ma, X., Guo, X., Fu, M., Intermetallics, 98, 2018, 11-17.

[39] Deng, H., Dickey, E.C., Paderno, Y., Paderno, V., Filippov, V., Journal of the American Ceramic Society, 90[8] 2007, 2603-2609.

[40] Yu, R., He, L.L., Guo, J.T., Ye, H.Q., Lupinc, V., Acta Materialia, 48[14] 2000, 3701-3710.

[41] Winkelman, G.B., Raviprasad, K., Muddle, B.C., Acta Materialia, 55[9] 2007, 3213-3228.

[42] Zhang, W.Z., Weatherly, G.C., Progress in Materials Science, 50[2] 2005, 181-292.

[43] Xu, W.S., Yang, X.P., Zhang, W.Z., Acta Metallurgica Sinica, 31[2] 2018, 113-126.

[44] Bollmann, W., Surface Science, 31, 1972, 1-11.

[45] Bonnet, R., Durand, F., Materials Research Bulletin, 7[10] 1972, 1045-1059.

[46] Smith, D.A., Pond, R.C., International Metals Reviews, 21[1] 1976, 61-74.

[47] Sadananda, K., Marcinkowski, M.J., Journal of Applied Physics, 45[4] 1974, 1521-1532.

[48] Sadananda, K., Marcinkowski, M.J., Scripta Metallurgica, 7[6] 1973, 557-563.

[49] Bollmann, W., Scripta Metallurgica, 7[6] 1973, 565-568.

[50] Mourey, M., Dabosi, F., Materials Research Bulletin, 9[4] 1974, 379-389.

[51] Mou, Y., Aaronson, H.I., Acta Metallurgica et Materialia, 42[6] 1994, 2133-2144.

[52] Bollmann, W., Poerry, A.J., Philosophical Magazine, 20[163] 1969, 33-50.

[53] Warrington, D.H., Bollmann, W., Philosophical Magazine, 25[5] 1972, 1195-1199.

[54] Shin, K., King, A.H., Materials Science and Engineering A, 113, 1989, 121-127.

[55] Rodríguez, A.G., Casajuana, D.R., Revista Mexicana de Fisica, 53[2] 2007, 139-143.

[56] Zhang, W.Z., Yang, X.P., Journal of Materials Science, 46[12] 2011, 4135-4156.

[57] Chen, H., Yang, Z., Acta Metallurgica Sinica, 43[7] 2007, 710-712.

[58] Ecob, R.C., Physica Status Solidi A, 71[2] 1982, 399-407.

[59] Zhang, J.Y., Gao, Y., Wang, Y., Zhang, W.Z., Acta Materialia, 165, 2019, 508-519.

[60] Dahmen, U., Scripta Metallurgica, 15[1] 1981, 77-81.

[61] Kato, M., Materials Transactions, JIM, 33[2] 1992, 89-96.

[62] Kato, M., Materials Science and Engineering A, 146[1-2] 1991, 205-216.

[63] Yuryev, D.V., Demkowicz, M.J., Applied Physics Letters, 105[22] 2014, 221601.

[64] Zhang, W.Z., Metallurgical and Materials Transactions A, 44[10] 2013, 4513-4531.

[65] Romeu, D., Gómez-Rodríguez, A., Acta Crystallographica A, 62[5] 2006, 411-412.

[66] Zhang, W.Z., Qiu, D., Yang, X.P., Ye, F., Metallurgical and Materials Transactions A, 37[3] 2006, 911-927.

[67] Tsurekawa, S., Morita, K., Nakashima, H., Yoshinaga, H., Materials Transactions, 38[5] 1997, 393-400.

[68] Goux, C., Canadian Metallurgical Quarterly, 13[1] 1974, 9-31.

[69] Lojkowski, W., Materials Research Society Symposium - Proceedings, 357, 1995, 407-412.

[70] Sangghaleh, A., Demkowicz, M.J., Computational Materials Science, 145, 2018, 35-47.

[71] Bollmann, W., Michaut, B., Sainfort, G., Physica Status Solidi A, 13[2] 1972, 637-649.

[72] Luo, C.P., Weatherly, G.C., Acta Metallurgica, 35[8] 1987, 1963-1972.

[73] Chen, Fu-Rong, King, A.H., Metallurgical Transactions. A, 19[9] 1988, 2359-2363.

[74] Cherkashin, N., Kononchuk, O., Hÿtch, M., Solid State Phenomena, 178-179, 2011, 489-494.

[75] Cherkashin, N., Kononchuk, O., Reboh, S., Hÿtch, M., Acta Materialia, 60[3] 2012, 1161-1173.

[76] Vaudin, M.D., Rühle, M., Sass, S.L., Acta Metallurgica, 31[7] 1983, 1109-1116.

[77] Eastman, J., Schmueckle, F., Vaudin, M.D., Sass, S.L., Advances in Ceramics, 10, 1984, 324-346.

[78] Zhu, Y., Philosophical Magazine A, 69[4] 1994, 717-728.

[79] Shashkov, D.A., Chisholm, M.F., Seidman, D.N., Acta Materialia, 47[15] 1999, 3939-3951.

[80] Ramamurthy, S., Carter, C.B., Physica Status Solidi A, 166[1] 1998, 37-55.

[81] Bartholomeusz, B.J., Lu, T.M., Rajan, K., Journal of Electronic Materials, 20[7] 1991, 759-765.

[82] Ramamoorthy, K., Sanjeeviraja, C., Jayachandran, M., Sankaranarayanan, K., Misra, P., Kukreja, L.M., Materials Chemistry and Physics, 84[1] 2004, 14-19.

[83] Zur, A., McGill, T.C., Journal of Applied Physics. 55[2] 1984, 378-386.

[84] Büschel, M., Tempel, A., Zehe, A., Crystal Research and Technology, 26[2] 1991, 211-215.

[85] Myers, T.H., Lo, Y., Bicknell, R.N., Schetzina, J.F., Applied Physics Letters, 42[3], 1983, 247-248.

[86] Raclariu, A.M., Deshpande, S., Bruggemann, J., Zhuge, W., Yu, T.H., Ratsch, C., Shankar, S., Computational Materials Science, 108, 2015, 88-93.

[87] Mathew, K., Singh, A.K., Gabriel, J.J., Choudhary, K., Sinnott, S.B., Davydov, A.V., Tavazza, F., Hennig, R.G., Computational Materials Science, 122, 2016, 183-190.

[88] Efimov, A.N., Lebedev, A.O., Thin Solid Films, 260[1] 1995, 111-117.

[89] May, B.J., Anderson, P.M., Myers, R.C., Journal of Crystal Growth, 459, 2017, 209-214.

[90] Runnels, B., Beyerlein, I.J., Conti, S., Ortiz, M., Journal of the Mechanics and Physics of Solids, 89, 2016, 174-193.

[91] Jelver, L., Larsen, P.M., Stradi, D., Stokbro, K., Jacobsen, K.W., Physical Review B, 96, 2017, 085306.

[92] Efimov, A.N., Lebedev, A.O., Surface Science, 344[3] 1995, 276-282.

[93] Lazić, P., Computer Physics Communications, 197, 2015, 324–334.

[94] Kawahara, K., Arafune, R., Kawai, M., Takagi, N., e-Journal of Surface Science and Nanotechnology, 13, 2015, 361-365.

[95] Zur, A., McGill, T.C., Nicolet, M.A., Journal of Applied Physics, 57[2] 1985, 600-603.

[96] Croke, E.T., Hauenstein, R.J., Nieh, C.W., McGill, T.C., Journal of Electronic Materials, 18[6] 1989, 757-761.

[97] Croke, E.T., Hauenstein, R.J., McGill, T.C., Applied Physics Letters, 53[6] 1988, 514-516.

[98] Dakshinamurthy, S., Rajan, K., JOM, 46[3] 1994, 52-53.

[99] Bulle-Lieuwma, C.W.T., Van Ommen, A.H., Hornstra, J., Aussems, C.N.A.M., Journal of Applied Physics, 71[5] 1992, 2211-2224.

[100] Lin, W.T., Wu, K.C., Pan, F.M., Thin Solid Films, 215[2] 1992, 184-187.

[101] Kang, T.S., Je, J.H., Kim, G.B., Baik, H.K., Lee, S.M., Journal of Vacuum Science and Technology B, 18[4] 2000, 1953-1956.

[102] Smeets, D., Vantomme, A., De Keyser, K., Detavernier, C., Lavoie, C., Journal of Applied Physics, 103[6] 2008, 063506.

[103] Shiau, F.Y., Cheng, H.C., Chen, L.J., Journal of Applied Physics, 59[8] 1986, 2784-2787.

[104] Heck, C., Kusaka, M., Hirai, M., Iwami, M., Yokota, Y., Thin Solid Films, 281-282[1-2] 1996, 94-97.

[105] Filonenko, O., Mogilatenko, A., Hortenbach, H., Allenstein, F., Beddies, G., Hinneberg, H.J., Journal of Crystal Growth, 262[1-4] 2004, 281-286.

[106] Van An, F., Bulenkov, N.A., Andreeva, A.V., Physica Status Solidi A, 88[2] 1985, 429-441.

[107] Dascalu, M., Cesura, F., Levi, G., Diéguez, O., Kohn, A., Goldfarb, I., Applied Surface Science, 476, 2019, 189-197.

[108] Geib, K.M., Mahan, J.E., Long, R.G., Nathan, M., Bai, G., Journal of Applied Physics, 70[3] 1991, 1730-1736.

[109] Mahan, J.E., Geib, K.M., Robinson, G.Y., Long, R.G., Xinghua, Y., Bai, G., Nicolet, M.A., Nathan, M., Applied Physics Letters, 56[21] 1990, 2126-2128.

[110] Catana, A., Schmid, P.E., Heintze, M., Lévy, F., Stadelmann, P., Bonnet, R., Journal of Applied Physics, 67[4] 1990, 1820-1825.

[111] Xie, W., Lucking, M., Chen, L., Bhat, I., Wang, G.C., Lu, T.M., Zhang, S., Crystal Growth and Design, 16[4] 2016, 2328-2334.

[112] Kim, K.H., Lee, J.J., Seo, D.J., Choi, C.K., Hong, S.R., Koh, J.D., Kim, S.C., Lee, J.Y., Nicolet, M.A., Journal of Applied Physics, 71[8] 1992, 3812-3815.

[113] Huang, J.Y., Wu, S.T., Japanese Journal of Applied Physics - 1, 38[6A] 1999, 3660-3663.

[114] He, Z., Stevens, M., Smith, D.J., Bennett, P.A., Surface Science, 524[1-3] 2003, 148-156.

[115] Goldfarb, I., Grossman, S., Cohen-Taguri, G., Applied Surface Science, 252[15] 2006, 5355-5360.

[116] Lin, W.T., Chen, L.J., Journal of Applied Physics, 59[5] 1986, 1518-1524.

[117] Chu, J.J., Chen, L.J., Tu, K.N., Journal of Applied Physics, 63[4] 1988, 1163-1167.

[118] Mahan, J.E., Geib, K.M., Robinson, G.Y., Long, R.G., Xinghua, Y., Bai, G., Nicolet, M.A., Nathan, M., Applied Physics Letters, 56[24] 1990, 2439-2441.

[119] Lee, Y.K., Fujimura, N., Ito, T., Itoh, N., Nanostructured Materials, 2[6] 1993, 603-614.

[120] Lin, W.T., Chen, L.J., Journal of Applied Physics, 59[10] 1986, 3481-3488.

[121] Chen, J.F., Chen, L.J., Thin Solid Films, 261[1-2] 1995, 107-114.

[122] Kawarada, H., Ishida, M., Nakanishi, J., Ohdomari, I., Horiuchi, S., Philosophical Magazine A, 54[5] 1986, 729-741.

[123] Konuma, K., Utsumi, H., Journal of Applied Physics, 76[4] 1994, 2181-2184.

[124] Kavanagh, K.L., Reuter, M.C., Tromp, R.M., Journal of Crystal Growth, 173[3-4] 1997, 393-401.

[125] Du, Y., Chen, K.H., Schuster, J.C., Perring, L., Huang, B.Y., Yuan, Z.H., Gachon, J.C., Zeitschrift für Metallkunde, 92[4] 2001, 323-327.

[126] Chang, Y.S., Chou, M.L., Journal of Applied Physics, 68[5] 1990, 2411-2414.

[127] Chang, Y.S., Chu, J.J., Materials Letters, 5[3] 1987, 67-71.

[128] Chang, C.S., Nieh, C.W., Chu, J.J., Chen, L.J., Thin Solid Films, 161, 1988, 263-271.

[129] Chen, Q., Xie, Q., Physics Procedia, 11, 2011, 134-137.

[130] Xiao, Q., Xie, Q., Shen, X., Zhang, J., Yu, Z., Zhao, K., Applied Surface Science, 257[17] 2011, 7800-7804.

[131] Chen, J.C., Shen, G.H., Chen, L.J., Journal of Applied Physics, 84[11] 1998, 6083-6087.

[132] Liu, B.Z., Nogami, J., Nanotechnology, 14[8] 2003, 873-877.

[133] Lee, Y.K., Fujimura, N., Ito, T., Itoh, N., Journal of Crystal Growth, 134[3-4] 1993, 247-254.

[134] Lee, Y.K., Lee, M.S., Lee, J.S., Journal of Crystal Growth, 244[3-4] 2002, 305-312.

[135] Frangis, N., Van Landuyt, J., Kaltsas, G., Travlos, A., Nassiopoulos, A.G., Journal of Crystal Growth, 172[1-2] 1997, 175-182.

[136] Yang, Y., Abelson, J.R., Journal of Crystal Growth, 310[13] 2008, 3197-3202.

[137] Li, B.Q., Zuo, J.M., Surface Science, 520[1-2] 2002, 7-17.
[138] Jin, H.S., Park, K.H., Yapsir, A.S., Wang, G.C., Lu, T.M., Luo, L., Gibson, W.M., Yamada, I., Takagi, T., Nuclear Instruments and Methods in Physics Research B, 40-41[2] 1989, 817-822.
[139] Jin, H.S., Yapsir, A.S., Lu, T.M., Gibson, W.M., Yamada, I., Takagi, T., Applied Physics Letters, 50[16] 1987, 1062-1064.
[140] Legoues, F.K., Liehr, M., Renier, M., Krakow, W., Philosophical Magazine B, 57[2] 1988, 179-189.
[141] Goswami, D.K., Bhattacharjee, K., Satpati, B., Roy, S., Kuri, G., Satyam, P.V., Dev, B.N., Applied Surface Science, 253[23] 2007, 9142-9147.
[142] Nason, T.C., You, L., Lu, T.M., Journal of Applied Physics, 72[2] 1992, 466-470.
[143] Bording, J.K., Li, B.Q., Shi, Y.F., Zuo, J.M., Physical Review Letters, 90[22] 2003, 226104.
[144] Naik, R., Kota, C., Rao, B.U.M., Journal of Vacuum Science and Technology A, 12[4] 1994, 1832-1837.
[145] Kato, M., Niwa, H., Philosophical Magazine B, 64[3] 1991, 317-326.
[146] Yapsir, A.S., Choi, C.H., Lu, T.M., Journal of Applied Physics, 67[2] 1990, 796-799.
[147] Liu, H., Zhang, Y.F., Wang, D.Y., Pan, M.H., Jia, J.F., Xue, Q.K., Surface Science, 571[1-3] 2004, 5-11.
[148] Liu, H., Zhang, Y.F., Wang, D.Y., Jia, J.F., Xue, Q.K., Chinese Physics Letters, 21[8] 2004, 1608-1611.
[149] Hasan, M.A., Radnoczi, G., Sundgren, J.E., Hansson, G.V., Surface Science, 236[1-2] 1990, 53-76.
[150] Westmacott, K.H., Hinderberger, S., Dahmen, U., Philosophical Magazine A, 81[6] 2001, 1547-1578.
[151] Hsieh, Y.F., Chen, L.J., Marshall, E.D., Lau, S.S., Thin Solid Films, 162, 1988, 287-294.
[152] Roy, A., Bhattacharjee, K., Dev, B.N., Applied Surface Science, 256[2] 2009, 508-512.
[153] Horn-von Hoegen, M., Henzler, M., Physica Status Solidi A, 146[1] 1994, 337-352.
[154] Ernst, F., Philosophical Magazine A, 68[6] 1993, 1251-1272.
[155] Ernst, F., Materials Research Society Symposium - Proceedings, 319, 1994, 165-170.
[156] Cao, S.P., Ye, F., Xu, A.Y., Bai, F.M., Materials Research Innovations, 18, 2014, S4642-S4645.
[157] Shinkai, S., Sasaki, K., Japanese Journal of Applied Physics - 1, 38[6A] 1999, 3646-3650.
[158] Oishi, N., Yanagisawa, H., Sasaki, K., Abe, Y., Kawamura, M., Electronics and Communications in Japan - II, 81[9] 1998, 46-52.
[159] Kaushik, V.S., Datye, A.K., Kendall, D.L., Martinez-Tovar, B., Myers, D.R., Applied Physics Letters, 52[21] 1988, 1782-1784.
[160] Kutana, A., Erwin, S.C., Physical Review B, 87[4] 2013, 045314.
[161] Jnawali, G., Hattab, H., Meyer Zu Heringdorf, F.J., Krenzer, B., Horn-Von Hoegen, M., Physical Review B, 76[3] 2007, 035337.
[162] Furdyna, J.K., Kossut, J., Superlattices and Microstructures, 2[1] 1986, 89-96.
[163] Cunningham, J.E., Pathak, R.N., Jan, W.Y., Applied Physics Letters, 68[3] 1996, 394-396.
[164] Jones, K.A., Tu, C.W., Journal of Crystal Growth, 70[1-2] 1984, 127-132.
[165] Allovon, M., Primot, J., Gao, Y., Quillec, M., Journal of Electronic Materials, 18[4] 1989, 505-510.
[166] Sagalowicz, L., Jouneau, P.H., Rudra, A., Syrbu, A.V., Kapon, E., Proceedings of the TMS Fall Meeting, 1998, 259-268.

[167] Xu, B., He, P., Liu, H., Wang, P., Zhou, G., Wang, X., Angewandte Chemie, 53[9] 2014, 2339-2343.

[168] Kozlovsky, V.I., Martovitsky, V.P., Skasyrsky, Y.K., Sadofyev, Y.G., Turyansky, A.G., Physica Status Solidi B, 229[1] 2002, 63-67.

[169] Tobin, S.P., Smith, F.T.J., Norton, P.W., Wu, J., Dudley, M., Marzio, D.D., Casagrande, L.G., Journal of Electronic Materials, 24[9] 1995, 1189-1199.

[170] Maekawa, T., Saito, T., Yoshikawa, M., Takigawa, H., Materials Research Society Symposia Proceedings, 56, 1986, 109-113.

[171] Uemoto, T., Kamata, A., Mitsuhashi, H., Hirahara, K., Beppu, T., Journal of Crystal Growth, 99[1-4] 1990, 422-426.

[172] Fujita, S., Terada, K., Sakamoto, T., Fujita, S., Journal of Crystal Growth, 94[1] 1989, 102-108.

[173] Ohmi, K., Suemune, I., Kanda, T., Kan, Y., Yamanishi, M., Japanese Journal of Applied Physics, 26[12A] 1987, L2072-L2075.

[174] Matsumura, N., Ishikawa, K., Saraie, J., Yodogawa, Y., Journal of Crystal Growth, 72[1-2] 1985, 41-45.

[175] Funato, M., Kitani, H., Fujita, S., Fujita, S., Journal of Electronic Materials, 25[2] 1996, 217-222.

[176] Kečkéš, J., Ortner, B., Červeň, I., Jakabovič, J., Kováč, J., Journal of Applied Physics, 80[11] 1996, 6204-6210.

[177] Sadowski, J., Herman, M.A., Journal of Crystal Growth, 146[1-4] 1995, 449-454.

[178] Rotter, S., Kasemset, D., Fonstad, C.G., IEEE Electron Device Letters, 3[3] 1982, 66-68.

[179] Shigenaka, K., Sugiura, L., Nakata, F., Hirahara, K., Journal of Crystal Growth, 145[1-4] 1994, 376-381.

[180] Ivanov, I.S., Sidorov, Y.G., Yakushev, M.V., Inorganic Materials, 33[3] 1997, 243-247.

[181] Sakurai, T., Matsumoto, N., Okada, Y., Onari, S., Akimoto, K., Physica Status Solidi C, 2[7] 2005, 2224-2227.

[182] Beanland, R., Kiely, C.J., Interface Science, 1[2] 1993, 99-113.

[183] Hung, L.S., Zheng, L.R., Blanton, T.N., Applied Physics Letters, 60[25] 1992, 3129-3131.

[184] Rajan, K., Gong, R., Webb, J., Applied Physics Letters, 57[14] 1990, 1446-1448.

[185] Ortner, B., Bauer, G., Journal of Crystal Growth, 92[1-2] 1988, 69-76.

[186] Shiau, F.Y., Chang, Y.A., Chen, L.J., Journal of Electronic Materials, 17[5] 1988, 433-441.

[187] Kenty, J.L., Thin Solid Films, 26[1] 1975, 181-195.

[188] Golding, T.D., Dura, J.A., Wang, W.C., Zborowski, J.T., Vigliante, A., Chen, H.C., Meyer, J.R., Journal of Crystal Growth, 127[1-4] 1993, 777-782.

[189] Golding, T.D., Dura, J.A., Wang, H., Zborowski, J.T., Vigliante, A., Chen, H.C., Miller, J.H., Meyer, J.R., Semiconductor Science and Technology, 8[1S] 1993, S117-S120.

[190] Golding, T.D., Dura, J.A., Wang, W.C., Vigliante, A., Moss, S.C., Chen, H.C., Miller, J.H., Hoffman, C.A., Meyer, J.R., Applied Physics Letters, 63[8] 1993, 1098-1100.

[191] Hsieh, J.J., Finn, M.C., Rossi, J.A., Institute of Physics - Conference Series, 33, 1976, 37-44.

[192] Wright, P.D., Rezek, E.A., Holonyak, N., Stillman, G.E., Rossi, J.A., Groves, W.O., Applied Physics Letters, 31[1] 1977, 40-42.

[193] Wright, P.D., Rezek, E.A., Holonyak, N., Journal of Crystal Growth, 41[2] 1977, 254-261.

[194] Tamura, A., Oka, K., Inoue, M., Shirafuji, J., Inuishi, Y., Proceedings of the Conference on Solid State Devices, 1980, 479-448.

[195] Tamura, A., Oka, K., Inoue, M., Shirafuji, J., Inuishi, Y., Japanese Journal of Applied Physics, 19, 1980, 479-482.

[196] Efimov, A.N., Lebedev, A.O., Tsaregorodtsev, A.M., Journal of Applied Crystallography, 31[3] 1998, 461-473.

[197] Zhylik, A., Benediktovitch, A., Feranchuk, I., Inaba, K., Mikhalychev, A., Ulyanenkov, A., Journal of Applied Crystallography, 46[4] 2013, 919-925.

[198] Kato, K., Sasaki, K., Abe, Y., Japanese Journal of Applied Physics - 1, 45[4A] 2006, 2731-2735.

[199] Kato, K., Sasaki, K., Abe, Y., Japanese Journal of Applied Physics - 1, 45[9A] 2006, 7097-7099.

[200] Seong, W.K., Oh, S., Kang, W.N., Japanese Journal of Applied Physics, 51[8-1] 2012, 083101.

[201] Koyama, T., Chichibu, S.F., Journal of Applied Physics, 95[12] 2004, 7856-7861.

[202] Morita, K., Tsurekawa, S., Nakashima, H., Yoshinaga, H., Journal of the Japan Institute of Metals, 59[9] 1995, 881-888.

[203] Stock, S.R., Ahn, S.H., Cohen, J.B., Journal of the American Ceramic Society, 70[3] 1987, 125-132.

[204] Ding, H., Dwaraknath, S.S., Garten, L., Ndione, P., Ginley, D., Persson, K.A., ACS Applied Materials and Interfaces, 8, 2016, 13086-13093.

[205] Watanabe, T., Saito, K., Osada, M., Suzuki, T., Fujimoto, M., Yoshimoto, M., Sasaki, A., Liu, J., Kakihana, M., Funakubo, H., Materials Research Society Symposium - Proceedings, 748, 2003, 69-74.

[206] Watanabe, T., Funakubo, H., Saito, K., Suzuki, T., Fujimoto, M., Osada, M., Noguchi, Y., Miyayama, M., Applied Physics Letters, 81[9] 2002, 1660-1662.

[207] Vuorinen, S., Hoel, R.H., Thin Solid Films, 232[1] 1993, 73-82.

[208] Tang, Y., Dai, F., Gu, X., Wang, Z., Zhang, W., Physica E, 77, 2016, 97-101.

[209] Jia, C.L., Hojczyk, R., Faley, M., Poppe, U., Urban, K., Philosophical Magazine A, 79[4] 1999, 873-891.

[210] Chiba, K., Makino, S., Mukaida, M., Kusunoki, M., Ohshima, S., IEEE Transactions on Applied Superconductivity, 11[1] 2001, 2734-2737.

[211] Ito, W., Yoshida, Y., Mahajan, S., Morishita, T., Journal of Crystal Growth, 146[1-4] 1995, 655-658.

[212] Vignolle, C., Gervais, A., Physica Status Solidi A, 126[1] 1991, 197-203.

[213] Lee, S.T., Chen, S., Hung, L.S., Braunstein, G., Applied Physics Letters, 55[3] 1989, 286-288.

[214] Shi, L., Zhou, G., Jia, Y., Huang, Y., Wang, R.L., Wang, C.A., Yi, H.A., Li, H.C., Zhang, Y., Superconductor Science and Technology, 6[3] 1993, 191-194.

[215] Inoue, K., Saito, M., Wang, Z., Kotani, M., Ikuhara, Y., Materials Transactions, 56[3] 2015, 281-287.

[216] Hay, R.S., Acta Materialia, 55[3] 2007, 991-1007.

[217] Ferdeghini, C., Grassano, G., Bellingeri, E., Marrè, D., Ramadan, W., Ferrando, V., Beneduce, C., International Journal of Modern Physics B, 17[4-6 II] 2003, 824-829.

[218] Gorbenko, O.Y., Bosak, A.A., Journal of Crystal Growth, 186[1-2] 1998, 181-188.

[219] Turan, S., Knowles, K.M., Interface Science, 8[2] 2000, 279-294.

[220] Peng, Y., Han, G., Wang, D., Wang, K., Guo, Z., Yang, J., Yuan, W., International Journal of Hydrogen Energy, 42[21] 2017, 14409-14417.

[221] Xu, Z., Salvador, P., Kitchin, J.R., ACS Applied Materials and Interfaces, 9[4] 2017, 4106-4118.

[222] Ohno, T., Ii, S., Shibata, N., Matsunaga, K., Ikuhara, Y., Yamamoto, T., Journal of the Japan Institute of Metals, 69[11] 2005, 1004-1009.

[223] Ohno, T., Ii, S., Shibata, N., Matsunaga, K., Ikuhara, Y., Yamamoto, T., Materials Transactions, 45[7] 2004, 2117-2121.

[224] Bristow, J.K., Butler, K.T., Svane, K.L., Gale, J.D., Walsh, A., Journal of Materials Chemistry A, 5[13] 2017, 6226-6232.

[225] Ding, H., Dwaraknath, S.S., Garten, L., Ndione, P., Ginley, D., Persson, K.A., ACS Applied Materials and Interfaces, 8[20] 2016, 13086-13093.

[226] Liu, H., Cheng, X., Valanoor, N., ACS Applied Materials and Interfaces, 8[50] 2016, 34844-34853.

[227] Wu, H., Chen, O., Zhuang, J., Lynch, J., Lamontagne, D., Nagaoka, Y., Cao, Y.C., Journal of the American Chemical Society, 133[36] 2011, 14327-14337.

[228] Sakellari, D., Frangis, N., Polychroniadis, E.K., Physica E, 42[5] 2010, 1777-1780.

[229] Lau, Y.K.A., Chernak, D.J., Bierman, M.J., Jin, S., Journal of Materials Chemistry, 19[7] 2009, 934-940.

[230] Mohanty, D., Sun, X., Lu, Z., Washington, M., Wang, G.C., Lu, T.M., Bhat, I.B., Journal of Applied Physics, 124[17] 2018, 175301.

[231] Littlejohn, A.J., Xiang, Y., Rauch, E., Lu, T.M., Wang, G.C., Journal of Applied Physics, 122[18] 2017, 185305.

[232] Gwo, S., Wu, C.L., Shen, C.H., Lin, H.W., Chen, H.Y., Ahn, H., Proceedings of SPIE - The International Society for Optical Engineering, 6134, 2006, 61340L.

[233] Wu, C.L., Shen, C.H., Lin, H.W., Lee, H.M., Gwo, S., Applied Physics Letters, 87[24] 2005, 241916.

[234] Kamohara, T., Akiyama, M., Ueno, N., Kuwano, N., Ceramics International, 34[4] 2008, 985-989.

[235] Kamohara, T., Akiyama, M., Ueno, N., Nonaka, K., Kuwano, N., Applied Physics Letters, 89[7] 2006, 071919.

[236] Shinkai, S., Sasaki, K., Yanagisawa, H., Yoshio, A., Japanese Journal of Applied Physics - 1, 42[10] 2003, 6518-6522.

[237] Abe, Y., Kawamura, M., Sasaki, K., Japanese Journal of Applied Physics - 1, 41[11B] 2002, 6857-6861.

[238] Efimov, A.N., Lebedev, A.O., Crystallography Reports, 47[1] 2002, 135-144.

[239] Lee, J.J., Kang, K.Y., Park, Y.S., Yang, C.S., Kim, H.S., Klm, K.H., Kang, T.W., Park, S.H., Lee, J.Y., Japanese Journal of Applied Physics - 1, 38[11] 1999, 6487-6488.

[240] Lee, J.J., Park, Y.S., Yang, C.S., Kim, H.S., Kim, K.H., Kang, K.Y., Kang, T.W., Park, S.H., Lee, J.Y., Journal of Crystal Growth, 213[1] 2000, 33-39.

[241] Ito, A., Masumoto, H., Goto, T., Thin Solid Films, 517[19] 2009, 5616-5620.

[242] Sandström, P., Svedberg, E.B., Birch, J., Sundgren, J.E., Journal of Crystal Growth, 197[4] 1999, 849-857.

[243] Svedberg, E.B., Sandström, P., Sundgren, J.E., Greene, J.E., Madsen, L.D., Surface Science, 429[1] 1999, 206-216.

[244] McCaffrey, J.P., Svedberg, E.B., Phillips, J.R., Madsen, L.D., Journal of Crystal Growth, 200[3] 1999, 498-504.

[245] Shoup, S.S., Paranthaman, M., Goyal, A., Specht, E.D., Lee, D.F., Kroeger, D.M., Beach, D.B., Journal of the American Ceramic Society, 81[11] 1998, 3019-3021.

[246] Unal, O., Mitchell, T.E., Journal of Materials Research, 7[6] 1992, 1445-1454.

[247] Ren, S.Y., Dow, J.D., Applied Physics Letters, 69[2] 1996, 251-253.

[248] Wang, S., Xu, X., Luo, H., Cao, C., Song, X., Zhao, J., Zhang, J., Tang, C., RSC Advances, 8[34] 2018, 19279-19288.

[249] Foronda, H.M., Mazumder, B., Young, E.C., Laurent, M.A., Li, Y., DenBaars, S.P., Speck, J.S., Journal of Crystal Growth, 475, 2017, 127-135.

[250] González, J.A., Andrés, J.P., López Antón, R., De Toro, J.A., Normile, P.S., Muniz, P., Riveiro, J.M., Nogués, J., Chemistry of Materials, 29[12] 2017, 5200-5206.

[251] Luo, S., Wang, C., Zhang, S., Tu, R., Liu, S., Tang, X., Shen, Q., Chen, F., Zhang, L., Applied Physics Express, 5[8] 2012, 085801.

[252] Sakai, Y., Saito, S., Cohen, M.L., Physical Review B, 89[11] 2014, 115424.

[253] Li, Y., Xiong, H., Chen, G., Yan, Z., Ji, X., Xu, K., Miao, L., Proceedings - International Conference on Natural Computation, 2013, 6818090, 829-833.

[254] Yang, C., Chen, Z., Hu, J., Ren, Z., Lin, S., Materials Research Bulletin, 47[6] 2012, 1331-1334.

[255] Li, L.B., Chen, Z.M., Xie, L.F., Yang, C., Journal of Crystal Growth, 385, 2014, 111-114.

[256] Yang, C., Chen, Z., Li, L., Li, W., Hu, J., Lin, S., Solid State Communications, 152[2] 2012, 68-70.

[257] Aravazhi, S., Geskus, D., Van Dalfsen, K., Vázquez-Córdova, S.A., Grivas, C., Griebner, U., García-Blanco, S.M., Pollnau, M., Applied Physics B, 111[3] 2013, 433-446.

[258] Yin, Z., Zhang, P., Zhang, M.S., Applied Physics Letters, 68[16] 1996, 2303-2305.

[259] Bugakov, A.V., Ievlev, V.M., Physics, Chemistry and Mechanics of Surfaces, 10[12] 1995, 1457-1470.

[260] Homma, H., Yang, K.Y., Schuller, I.K., Physical Review B, 36[18] 1987, 9435-9438.

[261] Schuller, I.K., Superlattices and Microstructures, 4[4-5] 1988, 521-524.

[262] Kaneko, T., Imafuku, M., Yamamoto, R., Doyama, M., Transactions of the Japan Institute of Metals, Supplement, 27, 1986, 323-328.

[263] Sakata, T., Yasuda, H.Y., Umakoshi, Y., Acta Materialia, 51[6] 2003, 1561-1572.

[264] Xu, W.S., Zhang, W.Z., Philosophical Magazine, 98[1] 2018, 75-93.

[265] Guziewski, M., Coleman, S.P., Weinberger, C.R., Acta Materialia, 155, 2018, 1-11.

[266] Guziewski, M., Coleman, S.P., Weinberger, C.R., Acta Materialia, 119, 2016, 184-192.

[267] Ye, F., Zhang, W.Z., Qiu, D., Acta Materialia, 54[20] 2006, 5377-5384.

[268] Han, S.Z., Park, S.I., Huh, J.S., Lee, Z.H., Lee, H.M., Materials Science and Engineering A, 230[1-2] 1997, 100-106.

[269] Zhou, J.P., Zhao, D.S., Zheng, O., Wang, J.B., Xiong, D.X., Sun, Z.F., Gui, J.N., Wang, R.H., Micron, 40[8] 2009, 906-910.

[270] Li, Y.J., Zhang, W.Z., Marthinsen, K., Acta Materialia, 60[17] 2012, 5963-5974.

[271] Matsuda, Y., Sakamoto, K., Yahisa, Y., Hosoe, Y., Hosoda, H., Kitamoto, Y., Journal of Magnetism and Magnetic Materials, 469, 2019, 545-549.

[272] Douin, J., Dahmen, U., Westmacott, K.H., Philosophical Magazine B, 63[4] 1991, 867-890.

[273] Budai, J.D., Young, R.T., Chao, B.S., Applied Physics Letters, 62[15] 1993, 1836-1838.

[274] Knorr, D.B., Merchant, S.M., Biberger, M.A., Journal of Vacuum Science and Technology B, 16[5] 1998, 2734-2744.

[275] Fartash, A., Thin Solid Films, 323[1-2] 1998, 296-303.

[276] Je, J.H., You, H., Cullen, W.G., Maroni, V.A., Ma, B., Koritala, R.E., Rupich, M.W., Thieme, C.L.H., Physica C, 384[1-2] 2003, 54-60.

[277] Pantleon, K., Somers, M.A.J., Acta Materialia, 52[16] 2004, 4929-4940.

[278] Pantleon, K., Somers, M.A.J., Materials Science Forum, 495-497[2] 2005, 1455-1460.

[279] Lu, Z., Sun, X., Washington, M.A., Lu, T.M., Journal of Physics D, 51[9] 2018, 095301.

[280] Chen, J.K., Reynolds, W.T., Acta Materialia, 45[11] 1997, 4423-4430.

[281] Tsurekawa, S., Tanaka, T., Nakashima, H., Yoshinaga, H., Journal of the Japan Institute of Metals, 58[4] 1994, 377-381.

[282] Sasajima, Y., Yamamoto, R., Doyama, M., Transactions of the Japan Institute of Metals, Supplement, 27, 1986, 301-306.

[283] Seki, A., Kame, K., Journal of the Iron and Steel Institute of Japan, 77[7] 1991, 892-897.

[284] Seki, A., Kame, K., ISIJ International, 32[12] 1992, 1306-1310.

[285] Chen, J.K., Ross, T.W., Chen, G., Kikuchi, M., Reynolds, W.T., Metallurgical and Materials Transactions A, 25[12] 1994, 2639-2646.

[286] Yang, J., Yang, Z., Qiu, D., Zhang, W., Zhang, C., Bai, B., Fang, H., Acta Metallurgica Sinica, 41[3] 2005, 225-230.

[287] Zhang, W.Z., Weatherly, G.C., Acta Materialia, 46[6] 1998, 1837-1847.

[288] Zhang, W.Z., Wu, J., Materials Science and Engineering A, 438-440[S] 2006, 118-121.

[289] Morita, K., Uehara, M., Tsurekawa, S., Nakashima, H., Journal of the Japan Institute of Metals, 61[4] 1997, 251-260.

[290] Wang, W., Cai, C., Rohrer, G.S., Gu, X., Lin, Y., Chen, S., Dai, P., Materials Characterization, 144, 2018, 411-423.

[291] Shao, X.H., Jin, Q.Q., Zhou, Y.T., Yang, H.J., Zheng, S.J., Zhang, B., Chen, Q., Ma, X.L., Materialia, 6, 2019, 100287.

[292] Lanxner, M., Bauer, C.L., Transactions of the Japan Institute of Metals, Supplement, 27, 1986, 617-624.

[293] Onda, T., Piao, M., Bando, Y., Ichinose, H., Otsuka, K., Materials Transactions, JIM, 36[1] 1995, 23-29.

[294] Wang, Y.G., Zhang, Z., Yan, G.H., De Hosson, J.T.M., Journal of Materials Science, 37[12] 2002, 2511-2518.

[295] Meissner, M., Sojka, F., Matthes, L., Bechstedt, F., Feng, X., Müllen, K., Mannsfeld, S.C.B., Forker, R., Fritz, T., ACS Nano, 10[7] 2016, 6474-6483.

[296] Ward, M.D., ACS Nano, 10[7] 2016, 6424-6428.

[297] Simbrunner, C., Schwabegger, G., Resel, R., Dingemans, T., Quochi, F., Saba, M., Mura, A., Bongiovanni, G., Sitter, H., Crystal Growth and Design, 14[11] 2014, 5719-5728.

[298] Sun, X., Lu, Z., Xie, W., Wang, Y., Shi, J., Zhang, S., Washington, M.A., Lu, T.M., Applied Physics Letters, 110[15] 2017, 153104.

[299] Littlejohn, A.J., Lu, T.M., Zhang, L.H., Kisslinger, K., Wang, G.C., CrystEngComm, 18[15] 2016, 2757-2769.

[300] Liang, J.J., Kung, P.W.C., Journal of Materials Research, 17[7] 2002, 1686-1691.

[301] Wallace, S.K., Butler, K.T., Hinuma, Y., Walsh, A., Journal of Applied Physics, 125[5] 2019, 055703.

[302] Beyer, P., Breuer, T., Ndiaye, S., Zykov, A., Viertel, A., Gensler, M., Rabe, J.P., Hecht, S., Witte, G., Kowarik, S., ACS Applied Materials and Interfaces, 6[23] 2014, 21484-21493.

www.ingramcontent.com/pod-product-compliance
Lightning Source LLC
Chambersburg PA
CBHW061021220326
41597CB00017BB/2141